50 Leveled Math Problems

150 Problems Total

Author

Anne M. Collins, Ph.D.

LESLEY
UNIVERSITY

SHELL EDUCATION

Contributing Author and Consultant

Linda Dacey, Ed.D.
Professor, Lesley University

Consultants

Kaitlyn Aspell
Achievement Center for Mathematics
Lesley University

Stephen Yurek
Achievement Center for Mathematics
Lesley University

Publishing Credits

Dona Herweck Rice, *Editor-in-Chief;* Robin Erickson, *Production Director;* Lee Aucoin, *Creative Director;* Timothy J. Bradley, *Illustration Manager;* Sara Johnson, M.S.Ed., *Senior Editor;* Aubrie Nielsen, M.S.Ed., *Associate Education Editor;* Jennifer Kim, M.A.Ed., *Associate Education Editor;* Leah Quillian, *Assistant Editor;* Grace Alba, *Interior Layout Designer;* Corinne Burton, M.A.Ed., *Publisher*

Standards

© 2003 National Council of Teachers of Mathematics (NCTM)
© 2004 Mid-continent Research for Education and Learning (McREL)
© 2007 Teachers of English to Speakers of Other Languages, Inc. (TESOL)
© 2010 National Governors Association Center for Best Practices and Council of Chief State School Officers (CCSS)

Shell Education
5301 Oceanus Drive
Huntington Beach, CA 92649-1030
http://www.shelleducation.com
ISBN 978-1-4258-0778-8
© 2012 Shell Educational Publishing, Inc.

Table of Contents

Table of Contents (cont.)

Problem Solving in Mathematics Instruction

If you were a student in elementary school before the early 1980s, your education most likely paid little or no attention to mathematical problem solving. In fact, your exposure may have been limited to solving word problems at the end of a chapter that focused on one of the four operations. After a chapter on addition, for example, you solved problems that required you to add two numbers to find the answer. You knew this was the case, so you just picked out the two numbers from the problem and added them. Sometimes, but rarely, you were assigned problems that required you to choose whether to add, subtract, multiply, or divide. Many of your teachers dreaded lessons that contained such problems as they did not know how to help the many students who struggled.

If you went to elementary school in the later 1980s or in the 1990s, it may have been different. You may have learned about a four-step model of problem solving and perhaps you were introduced to different strategies for finding solutions. There may have been a separate chapter in your textbook that focused on problem solving and two-page lessons that focused on particular problem-solving strategies, such as guess and check. Attention was given to problems that required more than one computational step for their solution, and all the information necessary to solve the problems was not necessarily contained in the problem statements.

One would think that the ability of students to solve problems would improve greatly with these changes, but that has not been the case. Research provides little evidence that teaching problem solving in this isolated manner leads to success (Cai 2010). In fact, some would argue that valuable instructional time was lost exploring problems that did not match the mathematical goals of the curriculum. An example would be learning how to use logic tables to solve a problem that involved finding out who drank which drink and who wore which color shirt. Being able to use a diagram to organize information, to reason deductively, and to eliminate possibilities are all important problem-solving skills, but they should be applied to problems that are mathematically significant and interesting to students.

Today, leaders in mathematics education recommend teaching mathematics in a manner that integrates attention to concepts, skills, and mathematical reasoning. Referred to as *teaching through problem solving,* this approach suggests that problematic tasks serve as vehicles through which students acquire new mathematical concepts and skills (D'Ambrosio 2003). Students apply previous learning and gain new insights into mathematics as they wrestle with challenging tasks. This approach is quite different from introducing problems only after content has been learned.

Most recently, the *Common Core State Standards* listed the need to persevere in problem solving as the first of its Standards for Mathematical Practice (National Governors Association Center for Best Practices and Council of Chief State School Officers 2010):

> **Make sense of problems and persevere in solving them.**
> *Mathematically-proficient students start by explaining to themselves the meaning of a problem and looking for entry points to its solution. They analyze givens, constraints, relationships, and goals. They make conjectures about the form and meaning of the solution and plan a solution pathway rather than simply jumping into a solution attempt. They consider analogous problems,*

Problem Solving in Mathematics Instruction *(cont.)*

and try special cases and simpler forms of the original problem in order to gain insight into its solution. They monitor and evaluate their progress and change course if necessary. Older students might, depending on the context of the problem, transform algebraic expressions or change the viewing window on their graphing calculator to get the information they need. Mathematically-proficient students can explain correspondences between equations, verbal descriptions, tables, graphs, or draw diagrams of important features and relationships, graph data, and search for regularity or trends. Younger students might rely on using concrete objects or pictures to help conceptualize and solve a problem. Mathematically proficient students check their answers to problems using a different method, and they continually ask themselves, "Does this make sense?" They can understand the approaches of others to solving complex problems and identify correspondences between different approaches.

This sustained commitment to problem solving makes sense; it is the application of mathematical skills to real-life problems that makes learning mathematics so important. Unfortunately, we have not yet mastered the art of developing successful problem solvers. Students' performance in the United States on the 2009 Program for International Student Assessment (PISA), a test that evaluates 15-year-old students' mathematical literacy and ability to apply mathematics to real-life situations, suggests that we need to continue to improve our teaching of mathematical problem solving. According to data released late in 2010, students in the United States are below average (National Center for Educational Statistics 2010). Clearly, we need to address this lack of success.

Students do not have enough opportunities to solve challenging problems. Further, problems available to teachers are not designed to meet the individual needs of students. Additionally, teachers have few opportunities to learn how best to create, identify, and orchestrate problem-solving tasks. This unique series, *50 Leveled Math Problems*, is designed to address these concerns.

Understanding the Problem-Solving Process

George Polya is known as the father of problem solving. In his book *How to Solve It: A New Aspect of Mathematical Method* (1945), Polya provides a four-step model of problem solving that has been adopted in many classrooms: understanding the problem, making a plan, carrying out the plan, and looking back. In some elementary classrooms, this model has been shortened: understand, plan, do, and check. Unfortunately, this over-simplification ignores much of the richness of Polya's thinking.

Polya's conceptual model of the problem-solving process has been adapted for use at this level. Teachers are encouraged to view the four steps as interrelated, rather than only sequential, and to recognize that problem-solving strategies are useful at each stage of the problem-solving process. The model presented here gives greater emphasis to the importance of communicating and justifying one's thinking as well as to posing problems. Ways in which understanding is deepened throughout the problem-solving process is considered in each of the following steps:

Step 1: Understand the Problem

Students engage in the problem-solving process when they attempt to understand the problem, but understanding is not something that happens just in the beginning. At grade 6, students may be asked to restate the problem in their own words and then turn to a neighbor to summarize what they know and what they need to find out. The teacher may read the problem aloud when working with a small group of English language learners.

What is most important is that teachers do not teach students to rely on key words or show students "tricks" or "short-cuts" that are not built on conceptual understanding. Interpreting the language of mathematics is complex, and terms that are used in mathematics often have different everyday meanings. Note how a reliance on key words would lead to failure when solving the problem below. A student taught that *of* means *multiply* may multiply the fractions to find that one-half of the students (17 students) is the correct answer to the problem.

> *There are 18 boys and 16 girls in a classroom. Some are seated and some are standing. $\frac{2}{3}$ of all the boys and $\frac{3}{4}$ of all the girls are seated. How many students are seated?*

Step 2: Apply Strategies

Once students have a sense of the problem, they can begin to actively explore it. They may do so by applying one or more of the following strategies. Note that we have combined related actions within some of the strategies.

- Act it out or use manipulatives.
- Count, compute, or write an equation.
- Find information in a picture, list, table, graph, or diagram.
- Generalize a pattern.
- Guess and check or make an estimate.
- Organize information in a picture, list, table, graph, or diagram.
- Simplify the problem.
- Use logical reasoning.
- Work backward.

Understanding the Problem-Solving Process *(cont.)*

Step 2: Apply Strategies *(cont.)*

As students apply these strategies, they also deepen their understanding of the mathematics of the problem. As such, understanding develops throughout the problem-solving process. Consider the following problem requiring students to find the missing values:

In the following equations, the same shapes have the same values.

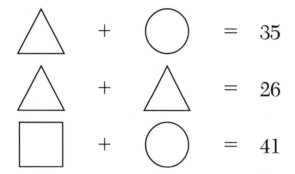

Kai and Gunnar are working together. They have read the problem and understand that they are to make the equations balance and find the values for the triangles, circles, and square. When Kai says, "We can just guess and check," Gunnar responds, "No, if we look at the two triangles, we know that 13 + 13 = 26, so we already know the value for the triangle." Kai answers, "I get it now. We can put the 13 in the first equation to find that the value of the circle is 35 – 13. And if the circle is worth 22, then the square must be worth 19."

The boys used logical reasoning and working backward to find the values and demonstrated an understanding of how the three equations are related.

It is important that we offer students problems that can be solved in more than one way. If one strategy does not lead to success, students can try a different one. This method gives students the opportunity to learn that getting "stuck" might just mean that a new approach should be considered. When students get themselves "unstuck," they are more likely to view themselves as successful problem solvers. Such problems also lead to richer mathematical conversations, as there are different ideas and perspectives to discuss. Consider the following problem:

Rick's box of cereal contains blue and purple marshmallows. He pours some cereal and sees 10 blue marshmallows and 8 purple marshmallows in his bowl. The next morning, he pours himself a much larger bowl of cereal. If the new bowl has 108 marshmallows that are in the same ratio of blue to purple, how many blue and purple marshmallows are in the larger bowl?

Understanding the Problem-Solving Process *(cont.)*

Step 2: Apply Strategies *(cont.)*

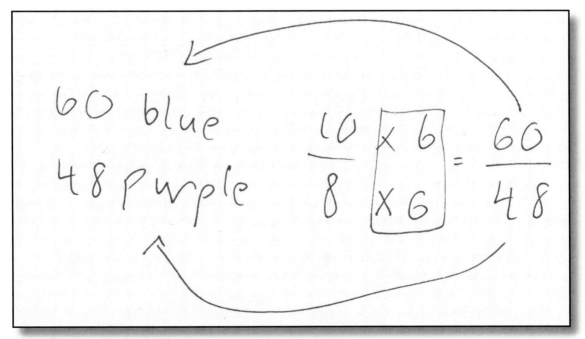

Student Sample 1

Notice that Sophia made a ratio table to solve this problem. She explained that she knew that both the blue and purple marshmallows had to increase at the same rate. She said that after each new ratio, she added them together to see if the sum was 108.

Student Sample 2

Jose used a scale factor. He explained that when he has to make an equivalent fraction, he needs to multiply by one because multiplying by one does not change its value, and multiplying by $\frac{6}{6}$ is the same as multiplying by one.

Understanding the Problem-Solving Process *(cont.)*

Step 3: Communicate and Justify Your Thinking

It is essential that teachers ask students to communicate and justify their thinking. It is also important that students make records as they work so that they can recall their thinking. When teachers make it clear that they expect such behavior from students, students are establishing an important habit of mind and developing their understanding of the nature of mathematics (Goldenberg, Shteingold and Feurzeig 2003). When students explain their thinking verbally while investigating a problem with a partner or small group, they deepen their understanding of the problem or may recognize an error and fix it. When students debrief after finding solutions, they learn to communicate their thinking clearly in ways that give others access to new mathematical ideas. In one class, a fourth-grader observed, "So, I think that if the same block is on each side of the scale, you can just cross them off. I mean, you know they are already equal." Such discourse is essential to the mathematical practice suggested in the *Common Core State Standards* that students "construct viable justifications and critique the reasoning of others" (National Governors Association Center for Best Practices and Council of Chief State School Officers 2010).

Our task is to foster learning environments where students engage in this kind of accountable talk. Michaels, O'Connor, and Resnick (2008) identify three aspects of this type of dialogue. The first is that students are accountable to their learning communities; they listen to each other carefully and build on the ideas of others. Second, accountable talk is based on logical thinking and leads to logical conclusions. Finally, these types of discussions are based on facts or other information that is available to everyone.

When we emphasize the importance of discussions and explanations, we are teaching our students that it is the soundness of their mathematical reasoning that determines what is correct, not merely an answer key or a teacher's approval. Therefore, students learn that mathematics makes sense and that they are mathematical sense-makers.

Understanding the Problem-Solving Process *(cont.)*

Step 4: Take It Further

Debrief

It is this final step in the problem-solving process to which teachers and students are most likely to give the least attention. When time is given to this step, it is often limited to *check your work.* In contrast, this step offers rich opportunities for further learning. Students might be asked to solve the problem using a different strategy or find additional solutions. They might be asked to make a mathematical generalization based on their investigation. Students might connect this problem to another problem they have solved already, or they now may be able to solve a new, higher-level problem.

Posing Problems

Students can also take problem solving further by posing problems. In fact, problem posing is intricately linked with problem solving (Brown and Walter 2005). When posing their own problems, students can view a problem as something they can create, rather than as a task that is given to them. This book supports problem posing through a variety of formats. For example, students may be asked to supply missing data in a problem so that it makes sense. They may be given a problem with the question omitted and asked to compose one. Or, they may be given both problem data and the answer and asked to identify the missing question. Teachers may also choose to ask students to create their own problems that are similar to those they have previously solved. Emphasis on problem posing can transform the teaching of problem solving and build lifelong curiosity in students.

Problem-Solving Strategies

Think of someone doing repair jobs around the house. Often that person carries a toolbox or wears a tool belt from task to task. Common tools, such as hammers, screwdrivers, and wrenches, are readily available. The repair person chooses tools (usually more than one) appropriate for a particular task. Problem-solving strategies are the tools used to solve problems. Labeling the strategies allows students to refer to them in discussions and helps students recognize the wide variety of tools available for the solution of problems. The problems in this book provide opportunities for students to apply one or more of the following strategies:

Act It Out or Use Manipulatives

Students' understanding of a problem is greatly enhanced when they act it out. Students may choose to dramatize a situation themselves or use manipulatives to show the actions or changes that take place. If students suggest they do not understand a problem, say something, such as *Imagine this is a play. Show me what the actors are doing.*

Count, Compute, or Write an Equation

When students count, compute, or write an equation to solve a problem, they are making a match between a context and a mathematical skill. Once the connection is made, students need only to carry out the procedure accurately. Sometimes writing an equation is a final step in the solution process. For example, students might work with manipulatives or draw pictures and then summarize their thinking by recording an equation.

Find Information in a Picture, List, Table, Graph, or Diagram

Too often problems contain all of the necessary information in the problem statement. Such information is never so readily available in real-world situations. It is important that students develop the ability to interpret a picture, list, table, graph, or diagram, and identify the information relevant to the problem.

Generalize a Pattern

Some people consider mathematics the study of patterns, so it makes sense that the ability to identify, continue, and generalize patterns is an important problem-solving strategy. The ability to generalize a pattern requires students to recognize and express relationships. Once generalized, the student can use the pattern to predict other outcomes.

Guess and Check or Make an Estimate

Guessing and checking or making an estimate provide students with insights into problems. Making a guess can help students to better understand conditions of the problem; it can be a way to try something when a student is stuck. Some students may make random guesses, but over time, students learn to make more informed guesses. For example, if a guess leads to an answer that is too large, a student might next try a number that is less than the previous guess. Estimation can help students narrow their range of guesses or be used to check a guess.

Problem-Solving Strategies (cont.)

Organize Information in a Picture, List, Table, Graph, or Diagram

Organizing information can help students both understand and solve problems. For example, students might draw a number line or a map to note information given in the problem statement. When students organize data in a table or graph, they might recognize relationships among the data. Students might also make an organized list to keep track of guesses they have made or to identify patterns. It is important that students gather data from a problem and organize it in a way that makes the most sense to them.

Simplify the Problem

Another way for students to better understand a problem, or perhaps get "unstuck," is to simplify it. Often the easiest way for students to do this is to make the numbers easier. For example, a student might replace four-digit numbers with single-digit numbers or replace fractions with whole numbers. With simpler numbers, students often gain insights or recognize relationships that were not previously apparent but that can now be applied to the original problem. Students might also work with 10 numbers, rather than 100, to identify patterns.

Use Logical Reasoning

Logical thinking and sense-making pervades mathematical problem solving. To solve problems, students need to deduce relationships, draw conclusions, make inferences, and eliminate possibilities. Logical reasoning is also a component of many other strategies. For example, students use logical reasoning to revise initial guesses or interpret diagrams. Asking questions, such as *What else does this sentence tell you?* helps students more closely analyze given data.

Work Backward

When the outcome of a situation is known, we often work backward to determine how to arrive at that goal. We might use this strategy to figure out what time to leave for the airport when we know the time our flight is scheduled to depart. A student might work backward to answer the question *What did Joey add to 79 to get a sum of 146?* or *If it took 2 hours and 23 minutes to drive a given, route and the driver arrived at 10:17, at what time did the driver leave home?* Understanding relationships among the operations is critical to the successful use of this strategy.

Ask, Don't Tell

All teachers want their students to succeed, and it can be difficult to watch them struggle. Often when students struggle with a problem, a first instinct may be to step in and show them how to solve it. That intervention might feel good, but it is not helpful to the student. Students need to learn how to struggle through the problem-solving process if they are to enhance their understanding and reasoning skills. Perseverance in solving problems is listed under the mathematical practices in the *Common Core State Standards* and research indicates that students who struggle and persevere in solving problems are more likely to internalize the problem-solving process and build upon their successes. It is also important to recognize the fact that people think differently about how to approach and solve problems.

An effective substitution for telling or showing students how to solve problems is to offer support through questioning. George Bright and Jeane Joyner (2005) identify three different types of questions to ask, depending on where students are in the problem-solving process: (1) engaging questions, (2) refocusing questions, and (3) clarifying questions.

Engaging Questions

Engaging questions are designed to pique a student's interest in a problem. Students are more likely to want to solve problems that are interesting and relevant. One way to immediately grab a student's attention is by using his or her name in the problem. Once a personal connection is made, a student is more apt to persevere in solving the problem. Posing an engaging question is also a great way to redirect a student who is not involved in a group discussion. Suppose students are provided the missing numbers in a problem and one of the sentences reads: *Janel is about _____ centimeters tall and rides her bicycle to school.* Engaging questions might include *What do you know about 100 centimeters? Are you taller or shorter than 100 centimeters?* The responses will provide further insight into how the student is thinking.

Refocusing Questions

Refocusing questions are asked to redirect students away from a nonproductive line of thinking and back to a more appropriate track. These questions often begin with the phrase *What can you tell me about…?* or *What does this number…?* Refocusing questions are also appropriate if you suspect students have misread or misunderstood the problem. Asking them to explain in their own words what the problem is stating and what question they are trying to answer is often helpful.

Clarifying Questions

Clarifying questions are posed when it is unclear why students have used a certain strategy, picture, table, graph, or computation. They are designed to help demonstrate what students are thinking, but can also be used to clear up misconceptions students might have. The teacher might say *I am not sure why you started with the number 10. Can you explain that to me?*

As teachers transform instruction from "teaching as telling" to "teaching as facilitating," students may require an adjustment period to become accustomed to the change in expectations. Over time, students will learn to take more responsibility and to expect the teacher to probe their thinking, rather than supply them with answers. After making this transition in her own teaching, one teacher shared a student's comment: "I know when I ask you a question that you are only going to ask me a question in response. But, sometimes the question helps me figure out the next step I need to take. I like that."

Differentiating with Leveled Problems

There are four main ways that teachers can differentiate: by content, by process, by product, and by learning environment. Differentiation by content involves varying the material that is presented to students. Differentiation by process occurs when a teacher delivers instruction to students in different ways. Differentiation by product asks students to present their work in different ways. Offering different learning environments, such as small group settings, is another method of differentiation. Students' learning styles, readiness levels, and interests determine which differentiation strategies are implemented. The leveled problems in this book vary aspects of mathematics problems so that students at various readiness levels can succeed. Mini-lessons include problems at three levels and ideas for differentiation. These are designated by the following symbols:

⬤   lower-level challenge

◻   on-level challenge

▲   above-level challenge

☆   English language learner support

Ideally, students solve problems that are at just the right level of challenge—beyond what would be too easy, but not so difficult as to cause extreme frustration (Sylwester 2003; Tomlinson 2003; Vygotsky 1986). The goal is to avoid both a lack of challenge, which might leave students bored, as well as too much of a challenge, which might lead to significant anxiety.

Differentiating with Leveled Problems *(cont.)*

There are a variety of ways to level problems. In this book, problems are leveled based on the concepts and skills required to find the solution. Problems are leveled by adjusting one or more of the following factors:

Complexity of the Mathematical Language

The mathematical language used in problems can have a significant impact on their level of challenge. For example, negative statements are more difficult to interpret than positive ones. *It is not an even number* is more complex than *It is an odd number.* Phrases, such as *at least* or *between,* also add to the complexity of the information. Further, words, such as *table, face,* and *plot,* can be challenging since their mathematical meaning differs from their everyday uses.

Complexity of the Task

There are various ways to change the complexity of the task. One example would be the number of solutions that students are expected to identify. Finding one solution that satisfies problem conditions is less challenging than finding more than one solution, which is even less difficult than identifying all possible solutions. Similarly, increases and decreases in the number of conditions that must be met and the number of steps that must be completed change the complexity of a problem.

Changing the Numbers

Sometimes it is the size of the numbers that is changed to differentiate the level of mathematical skills required. A problem may be more complex when it involves fractions, decimals, or negative integers. Sometimes changes to the "friendliness" of the numbers are made to adapt the difficulty level. For example, if two problems involve fractions, one with common denominators is simpler than one with unlike denominators.

Amount of Support

Some problems provide more support for learners than others. Providing a graphic organizer or a table that is partially completed is one way to provide added support for students. Offering information with pictures rather than words can also vary the level of support. The inclusion of such support often helps students better understand problems and may offer insights on how to proceed. The exclusion of support allows a student to take more responsibility for finding a solution, and it may make the task appear more abstract or challenging.

Management and Assessment

Differentiation Strategies for English Language Learners

Many English language learners may work at a high readiness level in many mathematical concepts, but may need support in accessing the language content. Specific suggestions for differentiating for English language learners can be found in the *Differentiate* section of the mini-lessons. Additionally, the strategies below may assist teachers in differentiating for English language learners.

- Allow students to draw pictures or provide oral responses as an alternative to written responses.

- Pose questions with question stems or frames. Example question stems/frames include: *What would happen if…?*, *Why do you think…?*, *How would you prove…?*, *How is _____ related to _____?*, and *Why is _____ important?*

- Use visuals to give context to questions. Add pictures or icons next to key words, or use realia to help students understand the scenario of the problem.

- Provide sentence stems or frames to help students articulate their thoughts. Sentence stems include: *This is important because…*, *This is similar because…*, and *This is different because…*. Sentence frames include: *I agree with _____ because…*, *I disagree with _____ because…*, and *I think _____ because…*.

- Partner English language learners with language-proficient students.

Management and Assessment *(cont.)*

Organization of the Mini-Lessons

The mini-lessons in this book are organized according to the domains identified in the *Common Core State Standards*, which have also been endorsed by the National Council of Teachers of Mathematics. At grade 6, these domains are *Ratios and Proportional Relationships*, *The Number System*, *Expressions and Equations*, *Geometry*, and *Statistics and Probability*. Though organized in this manner, the mini-lessons are independent of one another and may be taught in any order within a domain or among the domains. What is most important is that the lessons are implemented in the order that best fits a teacher's curriculum and practice.

Ways to Use the Mini-Lessons

There are a variety of ways to assign and use the mini-lessons, and they may be implemented in different lessons throughout the year. They can provide practice with new concepts or be used to maintain skills previously learned. The problems can be incorporated into a teacher's mathematics lessons once or twice each week, or they may be used to introduce extended or additional instructional periods. They can be used in the regular classroom with the whole class or in small groups. They can also be used to support Response to Intervention (RTI) and after-school programs.

It is important to remember that a student's ability to solve problems depends greatly on the specific content involved and may change over the course of the school year. Establish the expectation that problem assignment is flexible; sometimes students will be assigned to one level (circle, square, or triangle) and sometimes to another. On occasion, you may also wish to allow students to choose their own problems. Much can be learned from students' choices!

Students can also be assigned one, two, or all three of the problems to solve. Although leveled, some students who are capable of wrestling with complex problems need the opportunity to warm up first to build their confidence. Starting at a lower level serves these students well. Teachers may also find that students correctly assigned to a below- or on-level problem will be able to consider a problem at a higher level after solving one of the lower problems. Students can also revisit these problems, investigating those at the higher levels not previously explored.

Grouping Students to Solve Leveled Problems

A differentiated classroom often groups students in a variety of ways, based on the instructional goals of an activity or the tasks students must complete. At times, students may work in heterogeneous groups or pairs with students of varying readiness levels. Other activities may lend themselves to homogeneous groups or pairs of students who share similar readiness levels. Since the problems presented in this book provide below-level, on-level, and above-level challenges, you may wish to partner or group students with others who are working at the same readiness level.

Since students' readiness levels may vary for different mathematical concepts and change throughout a course of study, students may be assigned different levels of problems at different times. It is important that the grouping of students for solving leveled problems stays flexible. Struggling students who feel that they are constantly assigned to work with a certain partner or group may develop feelings of shame or stigma. Above-level students who are routinely assigned to the same group may become disinterested and cause behavior problems. Varying students' groups can help keep the activities engaging.

Management and Assessment *(cont.)*

Assessment for Learning

In recent years, increased attention has been given to summative assessment in schools. Significantly more instructional time is taken with weekly quizzes, chapter tests, and state-mandated assessments. These tests, although seen as tedious by many, provide information and reports about achievement to students, parents, administrators, and other interested stakeholders. However, these summative assessments often do not have a real impact on an individual student's learning. In fact, when teachers return quizzes and tests, many students look at the grade, and if it is "good," they bring the assessment home. If it is not an acceptable grade, they often just throw away the assessment.

Research shows that to have an impact on student learning, we should rely on assessments *for* learning, rather than on assessments *of* learning. That is, we should focus on assessment data we collect during the learning process, not after the instructional cycle is completed. These assessments for learning, or formative assessments, are shown to have the greatest positive impact on student achievement (National Mathematics Advisory Panel 2008). Assessment for learning is an ongoing process that includes a variety of strategies and protocols to inform the progression of students' learning.

One might ask, "So, what is the big difference? Don't all assessments accomplish the same goal?" The answer to those key questions is *no*. A great difference is the fact that formative assessment is designed to make student thinking visible. This is a real transformation for many teachers because when the emphasis is on students' thinking and reasoning, the focus shifts from whether the answer is correct or incorrect to how the students grapple with a problem. Making students' thinking visible entails a change in the manner in which teachers interact with their students. For instance, instead of relying solely on students' written work, teachers gather information through observation, questioning, and listening to their students discuss strategies, justify their reasoning, and explain why they chose to make particular decisions or use a specific representation. Since observations happen in real time, teachers can react in the moment by making an appropriate instructional decision, which may mean asking a well-posed question or suggesting a different model to represent the problem at hand.

Students are often asked to explain what they were thinking as they completed a procedure. Their response is often a recitation of the steps that were used. Such an explanation does not shed any light on whether a student understands the procedure, why it works, or if it will always work. Nor does it provide teachers with any insight into whether a student has a superficial or a deep understanding of the mathematics involved. If, however, students are encouraged to explain their thought processes, teachers will be able to discern the level of understanding. The vocabulary students use (or do not use) and the confidence with which they are able to answer probing questions can also provide insight into their levels of comprehension.

One of the most important features of formative assessment is that it actively involves students in their own learning. In assessment for learning, students are asked to reflect on their own work. They may be asked to consider multiple representations of a problem and then decide which of those representations makes the most sense, or which is the most efficient, or how they relate to one another. Students may be asked to make conjectures and then prove or disprove them by negation or counterexamples. Notice that it is the students doing the hard work of making decisions and thinking through the mathematical processes. Students who work at this level of mathematics, regardless of their grade level, demonstrate a deep understanding of mathematical concepts.

Management and Assessment (cont.)

Assessment for learning makes learning a shared endeavor between teachers and students. In effective learning environments, students take responsibility for their learning and feel safe taking risks, and teachers have opportunities to gain a deeper understanding of what their students know and are able to do. Implementing a variety of tools and protocols when assessing for learning can help the process become seamless. Some specific formative assessment tools and protocols include:

- Student Response Forms or Journals
- Range Questions
- Round-Robin Activities
- Gallery Walks
- Observation Protocols
- Feedback
- Exit Cards

Student Response Forms or Journals

Providing students with an organized workspace for the problems they solve can help a teacher to better understand a student's thinking and more easily identify misconceptions. Students often think that recording an answer is enough. If students do include further details, they often only write enough to fill the limited space that might be provided on an activity sheet. To promote the expectation that students show all of their work and record more of their thinking, use the included *Student Response Form* (page 132; studentresponse.pdf), or have students use a designated journal or notebook for solving problems. The prompts on the *Student Response Form* and the additional space provided encourage students to offer more details.

Range Questions

Range questions allow for a variety of responses, and teachers can use them to quickly gain access to students' understanding. Range questions are included in the activate section of many mini-lessons. The questions or problems that are posed are designed to provide insight into the spectrum of understanding that students bring to the day's problems. For instance, you might ask *What do you know about the numbers 43, 53, 63?* Students' responses may include stating that the factors of 63 are 3, 21, 9, and 7, or that 43 and 53 do not have factors other than 1 and the number itself, or that 43 and 53 are prime and 63 is composite. As you can imagine, the level of sophistication in the responses would vary and can help you decide which students to assign to which of the levels.

Round-Robin Activities

Round-robin activities are designed to facilitate teachers' abilities to see in real time how proficient students are with various mathematical procedures and concepts. Students are grouped in triads, and within each group, students are assigned the number 1, 2, or 3. The teacher announces that all number 1s go to the board. The teacher dictates an expression, equation, or computation for these students to record and complete one step. After completing the step, the number 1s sits down and the number 2s go to the board to do step 2. Upon completion of that step, these students return to their groups while the number 3s go to the board and do step 3. Students continue to go to the board in turn until the problem is complete. While the students are working at the board, the teacher has the opportunity to observe and provide individualized instruction and immediate feedback. At the same time, the format supports early intervention that prevents students from practicing, and thus reinforcing, any errors they make.

Management and Assessment (cont.)

Gallery Walks

Gallery walks can be used in many ways, but they all promote the sharing of students' problem-solving strategies and solutions. Pairs or small groups of students can record their pictures, tables, graphs, diagrams, computational procedures, and justifications on chart paper that they hang in designated areas of the classroom prior to the debriefing component of the lesson. Or, simply have students place their *Student Response Forms* at their workspaces and have students take a tour of their classmates' thinking. Though suggested occasionally for specific mini-lessons, you can include this strategy with any of the mini-lessons.

Observation Protocols

Observation protocols facilitate the data gathering that teachers must do as they document evidence of student learning. Assessment of learning is a key component in a teacher's ability to say, "I know that my students can apply these mathematical ideas because I have this evidence." Some important learning behaviors for teachers to focus on include: level of engagement in the problem/task; incorporation of multiple representations; inclusion of appropriate labels in pictures, tables, graphs, and solutions; use of accountable talk; inclusion of reflection on their work; and connections made between and among other mathematical ideas, previous problems, and their own life experiences. There is no one right form, nor could all of these areas be included on a form while leaving room for comments. Protocols should be flexible and allow teachers to identify categories of learning important to them and their students. An observation form is provided in the appendices (page 133; obs.pdf).

Feedback

Feedback is a critical component of formative assessment. Teachers who do not give letter grades on projects, quizzes, or tests, but who provide either neutral feedback or inquisitive feedback, find their students take a greater interest in the work they receive back than they did when their papers were graded. There are different types of feedback, but effective feedback focuses on the evidence in students' work. Many students respond favorably to an "assessment sandwich." The first comment might be a positive comment or praise for something well done, followed by a critical question or request for further clarification, followed by another neutral or positive comment.

Exit Cards

Exit cards are an effective way of assessing students' thinking at the end of a lesson in preparation for future instruction. There are multiple ways in which exit cards can be used. A similar problem to the one students have previously solved can be posed, or students can be asked to identify topics of confusion, what they liked best, or what they think they learned from a lesson. Some teachers use them to inform the following day's instruction. If students show more misconceptions than understanding, teachers use that information to add more practice with the concepts that caused the difficulty. Some exit card tasks are suggested in the *Differentiate* sections of the mini-lessons, but they may be added to any mini-lesson.

How to Use This Book

Mini-Lesson Plan

Lessons are organized by **Common Core State Standard** domains.

Suggested **Problem-Solving Strategies** outline strategies students may want to use in solving the problem. However, these are not the only strategies that can be used to solve the problem.

The McREL mathematics **Standards** for each lesson are provided.

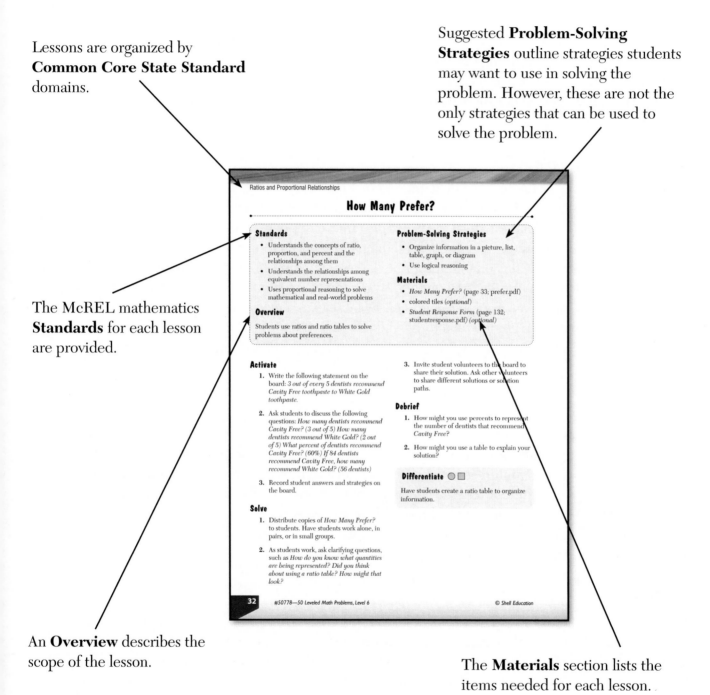

Ratios and Proportional Relationships

How Many Prefer?

Standards
- Understands the concepts of ratio, proportion, and percent and the relationships among them
- Understands the relationships among equivalent number representations
- Uses proportional reasoning to solve mathematical and real-world problems

Overview

Students use ratios and ratio tables to solve problems about preferences.

Problem-Solving Strategies
- Organize information in a picture, list, table, graph, or diagram
- Use logical reasoning

Materials
- *How Many Prefer?* (page 33; prefer.pdf)
- colored tiles *(optional)*
- *Student Response Form* (page 132; studentresponse.pdf) *(optional)*

Activate
1. Write the following statement on the board: *3 out of every 5 dentists recommend Cavity Free toothpaste to White Gold toothpaste.*
2. Ask students to discuss the following questions: *How many dentists recommend Cavity Free?* (3 out of 5) *How many dentists recommend White Gold?* (2 out of 5) *What percent of dentists recommend Cavity Free?* (60%) *If 84 dentists recommend Cavity Free, how many recommend White Gold?* (56 dentists)
3. Record student answers and strategies on the board.

Solve
1. Distribute copies of *How Many Prefer?* to students. Have students work alone, in pairs, or in small groups.
2. As students work, ask clarifying questions, such as *How do you know what quantities are being represented? Did you think about using a ratio table? How might that look?*

3. Invite student volunteers to the board to share their solution. Ask other volunteers to share different solutions or solution paths.

Debrief
1. How might you use percents to represent the number of dentists that recommend *Cavity Free?*
2. How might you use a table to explain your solution?

Differentiate ○ ■

Have students create a ratio table to organize information.

32   #50778—50 Leveled Math Problems, Level 6   © Shell Education

An **Overview** describes the scope of the lesson.

The **Materials** section lists the items needed for each lesson.

How to Use This Book (cont.)

Mini-Lesson Plan (cont.)

The **Activate** section suggests how you can access or assess students' prior knowledge. This section might recommend ways to have students review vocabulary, recall experiences related to the problem contexts, remember relevant mathematical ideas, or solve simpler related problems.

The **Solve** section provides suggestions on how to group students for the problem they will solve. It also provides questions to ask, observations to make, or procedures to follow to guide students in their work.

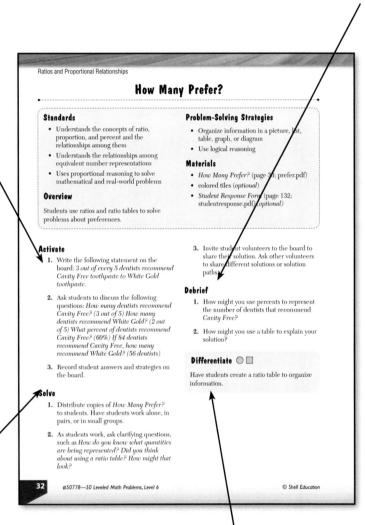

Ratios and Proportional Relationships

How Many Prefer?

Standards
- Understands the concepts of ratio, proportion, and percent and the relationships among them
- Understands the relationships among equivalent number representations
- Uses proportional reasoning to solve mathematical and real-world problems

Overview

Students use ratios and ratio tables to solve problems about preferences.

Problem-Solving Strategies
- Organize information in a picture, list, table, graph, or diagram
- Use logical reasoning

Materials
- *How Many Prefer?* (page 36; prefer.pdf)
- colored tiles (optional)
- *Student Response Form* (page 132; studentresponse.pdf) (optional)

Activate
1. Write the following statement on the board: *3 out of every 5 dentists recommend Cavity Free toothpaste to White Gold toothpaste.*
2. Ask students to discuss the following questions: *How many dentists recommend Cavity Free? (3 out of 5) How many dentists recommend White Gold? (2 out of 5) What percent of dentists recommend Cavity Free? (60%) If 84 dentists recommend Cavity Free, how many recommend White Gold? (56 dentists)*
3. Record student answers and strategies on the board.

Solve
1. Distribute copies of *How Many Prefer?* to students. Have students work alone, in pairs, or in small groups.
2. As students work, ask clarifying questions, such as *How do you know what quantities are being represented? Did you think about using a ratio table? How might that look?*

3. Invite student volunteers to the board to share their solution. Ask other volunteers to share different solutions or solution paths.

Debrief
1. How might you use percents to represent the number of dentists that recommend Cavity Free?
2. How might you use a table to explain your solution?

Differentiate ⦾ ▢

Have students create a ratio table to organize information.

32 #50778—50 Leveled Math Problems, Level 6   © Shell Education

The **Debrief** section provides questions designed to deepen students' understanding of the mathematics and the problem-solving process. Because the leveled problems share common features, it is possible to debrief either with small groups or as a whole class.

The **Differentiate** section includes additional suggestions to meet the unique needs of students. This section may offer support for English language learners, scaffolding for below-level students, or enrichment opportunities for above-level students. The following symbols are used to indicate appropriate readiness levels for the differentiation:

⦾ below level

▢ on level

△ above level

☆ English language learner

How to Use This Book *(cont.)*

Lesson Resources

Leveled Problems

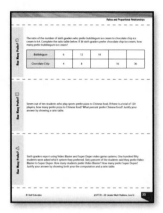

Each activity sheet offers **leveled problems** at three levels of challenge—below-level, on-level, and above-level. Cut the activity sheet apart and distribute the appropriate problem to each student, or present all of the leveled problems on an activity sheet to every student.

Record-Keeping Chart

Use the **Record-Keeping Chart** (page 134) to keep track of the problems each student completes.

Observation Form

Use the **Observation Form** (page 133) to document students' progress as they work through problems on their own and with their peers.

Teacher Resource CD

Helpful reproducibles are provided on the accompanying **Teacher Resource CD**. A detailed listing of the CD contents can be found on pages 141–143. The CD includes:

- Resources to support the implementation of the mini-lessons
- Manipulative templates
- Reproducible PDFs of all leveled problems and assessment tools
- Correlations to standards

How to Use This Book (cont.)

Lesson Resources (cont.)

Student Response Form

Students can attach their leveled problem to the form.

Students have space to show their work, provide their solution, and explain their thinking.

Appendix A

Name: _____   Date: _____

Student Response Form

Problem:

(glue your problem here)

My Work and Illustrations:
(picture, table, list, graph)

My Solution:

My Explanation:

_____

_____

_____

132   #50778—50 Leveled Math Problems, Level 6   © Shell Education

Correlations to Standards

Shell Education is committed to producing educational materials that are research- and standards-based. In this effort, we have correlated all of our products to the academic standards of all 50 United States, the District of Columbia, the Department of Defense Dependent Schools, and all Canadian provinces. We have also correlated to the *Common Core State Standards*.

How To Find Standards Correlations

To print a customized correlation report of this product for your state, visit our website at **http://www.shelleducation.com** and follow the on-screen directions. If you require assistance in printing correlation reports, please contact Customer Service at 1-877-777-3450.

Purpose and Intent of Standards

Legislation mandates that all states adopt academic standards that identify the skills students will learn in kindergarten through grade twelve. Many states also have standards for Pre-K. This same legislation sets requirements to ensure the standards are detailed and comprehensive.

Standards are designed to focus instruction and guide adoption of curricula. Standards are statements that describe the criteria necessary for students to meet specific academic goals. They define the knowledge, skills, and content students should acquire at each level. Standards are also used to develop standardized tests to evaluate students' academic progress. Teachers are required to demonstrate how their lessons meet state standards. State standards are used in the development of all of our products, so educators can be assured they meet the academic requirements of each state.

McREL Compendium

We use the Mid-continent Research for Education and Learning (McREL) Compendium to create standards correlations. Each year, McREL analyzes state standards and revises the compendium. By following this procedure, McREL is able to produce a general compilation of national standards. Each lesson in this product is based on one or more McREL standards.

TESOL Standards

The lessons in this book promote English language development for English language learners. The standards listed on the Teacher Resource CD (tesol.pdf) support the language objectives presented throughout the lessons.

Common Core State Standards

The lessons in this book are aligned to the Common Core State Standards (CCSS). The standards listed on pages 27–31 (ccss.pdf) support the objectives presented throughout the lessons.

NCTM Standards

The lessons in this book are aligned to the National Council of Teachers of Mathematics (NCTM) standards. The standards listed on the Teacher Resource CD (nctm.pdf) support the objectives presented throughout the lessons.

Correlations to Standards *(cont.)*

Common Core State Standards Correlation

<table>
<tr><th colspan="2">Common Core Standard</th><th>Lesson</th></tr>
<tr><td rowspan="6">Ratios and Proportional Relationships</td><td>**6.RP.1** Understand the concept of a ratio and use ratio language to describe a ratio relationship between two quantities.</td><td>How Many Prefer?, page 32; Mixing It Up, page 40</td></tr>
<tr><td>**6.RP.2** Understand the concept of a unit rate *a/b* associated with a ratio *a:b* with *b* ≠ 0, and use rate language in the context of a ratio relationship.</td><td>Best Buys, page 36</td></tr>
<tr><td>**6.RP.3** Use ratio and rate reasoning to solve real-world and mathematical problems, e.g., by reasoning about tables of equivalent ratios, tape diagrams, double number line diagrams, or equations.</td><td>How Many Prefer?, page 32; Survey Results, page 34; Best Buys, page 36; Paint Colors, page 38; Mixing It Up, page 40; Percent Tables, page 42</td></tr>
<tr><td>**6.RP.3a** Make tables of equivalent ratios relating quantities with whole-number measurements, find missing values in the tables, and plot the pairs of values on the coordinate plane. Use tables to compare ratios.</td><td>How Many Prefer?, page 32; Survey Results, page 34; Paint Colors, page 38; Mixing It Up, page 40</td></tr>
<tr><td>**6.RP.3b** Solve unit rate problems including those involving unit pricing and constant speed.</td><td>Best Buys, page 36; Going the Distance, page 90; How Far Did I Go?, page 100</td></tr>
<tr><td>**6.RP.3c** Find a percent of a quantity as a rate per 100; solve problems involving finding the whole, given a part and the percent.</td><td>Survey Results, page 34; Percent Tables, page 42</td></tr>
<tr><td rowspan="2">The Number System</td><td>**6.NS.1** Interpret and compute quotients of fractions, and solve word problems involving division of fractions by fractions, e.g., by using visual fraction models and equations to represent the problem.</td><td>How Many Groups?, page 44</td></tr>
<tr><td>**6.NS.3** Fluently add, subtract, multiply, and divide multi-digit decimals using the standard algorithm for each operation.</td><td>Garden Areas, page 66; Bank On It, page 70</td></tr>
</table>

Correlations to Standards (cont.)

Common Core State Standards Correlation (cont.)

Common Core Standard	Lesson
6.NS.4 Find the greatest common factor of two whole numbers less than or equal to 100 and the least common multiple of two whole numbers less than or equal to 12. Use the distributive property to express a sum of two whole numbers 1–100 with a common factor as a multiple of a sum of two whole numbers with no common factor.	Identical Groups, page 46; Are We in Sync?, page 48; Factors or Multiples?, page 50
6.NS.5 Understand that positive and negative numbers are used together to describe quantities having opposite directions or values; use positive and negative numbers to represent quantities in real-world contexts, explaining the meaning of 0 in each situation.	Greater or Less Than Zero?, page 54
6.NS.6 Understand a rational number as a point on the number line. Extend number line diagrams and coordinate axes familiar from previous grades to represent points on the line and in the plane with negative number coordinates.	Greater or Less Than Zero?, page 54
6.NS.6a Recognize opposite signs of numbers as indicating locations on opposite sides of 0 on the number line; recognize that the opposite of the opposite of a number is the number itself, e.g., –(–3) = 3, and that 0 is its own opposite.	Opposites Attract, page 58
6.NS.6b Understand signs of numbers in ordered pairs as indicating locations in quadrants of the coordinate plane; recognize that when two ordered pairs differ only by signs, the locations of the points are related by reflections across one or both axes.	Coordinate Graphing, page 60
6.NS.6c Find and position integers and other rational numbers on a horizontal or vertical number line diagram; find and position pairs of integers and other rational numbers on a coordinate plane.	Greater or Less Than Zero?, page 54; Integer Values, page 56; Opposites Attract, page 58
6.NS.7b Write, interpret, and explain statements of order for rational numbers in real-world contexts.	Integer Values, page 56
6.NS.8 Solve real-world and mathematical problems by graphing points in all four quadrants of the coordinate plane. Include use of coordinates and absolute value to find distances between points with the same first coordinate or the same second coordinate.	Coordinate Graphing, page 60; Polygons on the Plane, page 114

The Number System (cont.)

Correlations to Standards (cont.)

Common Core State Standards Correlation (cont.)

<table>
<tr><th colspan="2">Common Core Standard</th><th>Lesson</th></tr>
<tr><td rowspan="8" style="writing-mode: vertical-rl">Expressions and Equations</td><td>**6.EE.1** Write and evaluate numerical expressions involving whole-number exponents.</td><td>Exponentials, page 72</td></tr>
<tr><td>**6.EE.2a** Write expressions that record operations with numbers and with letters standing for numbers.</td><td>Expressly What?, page 96</td></tr>
<tr><td>**6.EE.2c** Evaluate expressions at specific values of their variables. Include expressions that arise from formulas used in real-world problems. Perform arithmetic operations, including those involving whole-number exponents, in the conventional order when there are no parentheses to specify a particular order (Order of Operations).</td><td>Evaluate Me, page 76; Simplify Me, page 98; Equivalences, page 102</td></tr>
<tr><td>**6.EE.5** Understand solving an equation or inequality as a process of answering a question: which values from a specified set, if any, make the equation or inequality true? Use substitution to determine whether a given number in a specified set makes an equation or inequality true.</td><td>Variable Value, page 78; Systems of Equations, page 80</td></tr>
<tr><td>**6.EE.6** Use variables to represent numbers and write expressions when solving a real-world or mathematical problem; understand that a variable can represent an unknown number, or, depending on the purpose at hand, any number in a specified set.</td><td>Arithmetic Sequences, page 88; Various and Sundry Patterns, page 94; Expressly What?, page 96</td></tr>
<tr><td>**6.EE.7** Solve real-world and mathematical problems by writing and solving equations of the form $x + p = q$ and $px = q$ for cases in which p, q and x are all nonnegative rational numbers.</td><td>Heads and Feet, page 82; My Equation Is, page 86</td></tr>
<tr><td>**6.EE.9** Use variables to represent two quantities in a real-world problem that change in relationship to one another; write an equation to express one quantity, thought of as the dependent variable, in terms of the other quantity, thought of as the independent variable. Analyze the relationship between the dependent and independent variables using graphs and tables, and relate these to the equation. For example, in a problem involving motion at constant speed, list and graph ordered pairs of distances and times, and write the equation $d = 65t$ to represent the relationship between distance and time.</td><td>Heads and Feet, page 82; My Equation Is, page 86</td></tr>
</table>

Correlations to Standards *(cont.)*

Common Core State Standards Correlation *(cont.)*

Common Core Standard	Lesson
6.G.1 Find the area of right triangles, other triangles, special quadrilaterals, and polygons by composing into rectangles or decomposing into triangles and other shapes; apply these techniques in the context of solving real-world and mathematical problems.	Quadrilaterals and Triangles, page 104; Boxy Areas, page 106; Dot Polygons, page 108
6.G.2 Find the volume of a right rectangular prism with fractional edge lengths by packing it with unit cubes of the appropriate unit fraction edge lengths, and show that the volume is the same as would be found by multiplying the edge lengths of the prism. Apply the formulas $V = lwh$ and $V = bh$ to find volumes of right rectangular prisms with fractional edge lengths in the context of solving real-world and mathematical problems.	Packaging Candy, page 112
6.G.3 Draw polygons in the coordinate plane given coordinates for the vertices; use coordinates to find the length of a side joining points with the same first coordinate or the same second coordinate. Apply these techniques in the context of solving real-world and mathematical problems.	Polygons on the Plane, page 114
6.G.4 Represent three-dimensional figures using nets made up of rectangles and triangles, and use the nets to find the surface area of these figures. Apply these techniques in the context of solving real-world and mathematical problems.	Nets, page 110

(The left margin of the table is labeled vertically: Geometry)

Correlations to Standards *(cont.)*

Common Core State Standards Correlation *(cont.)*

	Common Core Standard	Lesson
Statistics and Probability	**6.SP.1** Recognize a statistical question as one that anticipates variability in the data related to the question and accounts for it in the answers.	Statistical Questions, page 124
	6.SP.4 Display numerical data in plots on a number line, including dot plots, histograms, and box plots.	Stem-and-Leaf, page 126; Line Plots, page 128; Box and Whiskers, page 130
	6.SP.5c Giving quantitative measures of center (median and/or mean) and variability (interquartile range and/or mean absolute deviation), as well as describing any overall pattern and any striking deviations from the overall pattern with reference to the context in which the data were gathered.	Center It, page 118; Change It, page 120; Mean It, page 122; Line Plots, page 128; Box and Whiskers, page 130
	6.SP.5d Relating the choice of measures of center and variability to the shape of the data distribution and the context in which the data were gathered.	Center It, page 118

How Many Prefer?

Standards

- Understands the concepts of ratio, proportion, and percent and the relationships among them
- Understands the relationships among equivalent number representations
- Uses proportional reasoning to solve mathematical and real-world problems

Overview

Students use ratios and ratio tables to solve problems about preferences.

Problem-Solving Strategies

- Organize information in a picture, list, table, graph, or diagram
- Use logical reasoning

Materials

- *How Many Prefer?* (page 33; prefer.pdf)
- colored tiles (*optional*)
- *Student Response Form* (page 132; studentresponse.pdf) (*optional*)

Activate

1. Write the following statement on the board: *3 out of every 5 dentists recommend Cavity Free toothpaste to White Gold toothpaste.*

2. Ask students to discuss the following questions: *How many dentists recommend Cavity Free? (3 out of 5) How many dentists recommend White Gold? (2 out of 5) What percent of dentists recommend Cavity Free? (60%) If 84 dentists recommend Cavity Free, how many recommend White Gold? (56 dentists)*

3. Record student answers and strategies on the board.

Solve

1. Distribute copies of *How Many Prefer?* to students. Have students work alone, in pairs, or in small groups.

2. As students work, ask clarifying questions, such as *How do you know what quantities are being represented? Did you think about using a ratio table? How might that look?*

3. Invite student volunteers to the board to share their solution. Ask other volunteers to share different solutions or solution paths.

Debrief

1. How might you use percents to represent the number of dentists that recommend *Cavity Free*?

2. How might you use a table to explain your solution?

Differentiate ◓ ▢

Have students create a ratio table to organize information.

The ratio of the number of sixth graders who prefer bubblegum ice cream to chocolate chip ice cream is 6:4. Complete the ratio table below. If 36 sixth graders prefer chocolate chip ice cream, how many prefer bubblegum ice cream?

Bubblegum	6	12	18		
Chocolate Chip	4	8		16	36

Seven out of ten students who play sports prefer pizza to Chinese food. If there is a total of 120 players, how many prefer pizza to Chinese food? What percent prefer Chinese food? Justify your answer by showing a ratio table.

Sixth graders report using Video Blaster and Super Duper video game systems. One hundred fifty students were asked which system they preferred. Sixty percent of the students said they prefer Video Blaster to Super Duper. How many students prefer Video Blaster? How many prefer Super Duper? Justify your answer by showing both your the computation and a ratio table.

Survey Results

Standards

- Understands the concepts of ratio, proportion, and percent and the relationships among them
- Uses proportional reasoning to solve mathematical and real-world problems

Overview

Students use ratios and percents to solve problems based on survey results. They may use a bar diagram to represent the ratios

Problem-Solving Strategies

- Organize information in a picture, list, table, graph, or diagram
- Simplify the problem
- Use logical reasoning

Materials

- *Survey Results* (page 35; survey.pdf)
- snap cubes in two colors
- *Student Response Form* (page 132; studentresponse.pdf) *(optional)*

Activate

1. Distribute a handful of two different-colored snap cubes to each student. Have students work in pairs to model three representations of the ratio of 2:5. *(Possibilities include 4:10, 6:15, and 8:20.)*

2. Tell students that the result of a survey showed that 2 out of 5 people prefer the B-Box to the Woo game system.

3. Ask *How many people prefer the B-Box if 1,275 people were surveyed? (510)* Have students think independently before instructing them to turn to a partner to discuss a strategy they might use to answer the question.

4. Ask students to share their strategies. Have students complete the problem. Ask several volunteers to share their solutions and solution paths.

Solve

1. Distribute *Survey Results* to students. Have students work alone, in pairs, or in small groups.

2. Encourage students to model the problem using the snap cubes or a picture, or make a ratio table.

3. Invite several volunteers to share their solutions and solution paths.

Debrief

1. How might you use a diagram to explain your solution?

2. How might you use a table to explain your solution?

3. How might you break the problem into a simpler problem before solving it?

Differentiate ⚪ ◻

Have students use a bar model to show the part to whole relationship.

The sixth-grade basketball team took a survey of how many students in their school prefer basketball to soccer. Of the 1,740 students responding, 2 out of every 3 students said they prefer basketball. How many students prefer basketball? Complete the bar diagram to support your answer.

The Friendly Pet Store owners discovered that 6 out of 10 of their customers own dogs that prefer Doggie Delicious Donut treats to Crunchie Cookie treats. If they have 12,450 customers, how many dogs prefer Crunchie Cookie treats? How many prefer Doggie Delicious Donut treats? Draw a bar diagram to support your solution.

Dippy Donuts surveyed 17,850 customers to determine whether they prefer bagels or donuts. Of the 17,850 customers, 10,710 stated they prefer donuts. What is the ratio of customers who prefer donuts to customers who prefer bagels? Justify your response using a ratio table or a bar diagram.

Best Buys

Standards

- Understands the concepts of ratio, proportion, and percent and the relationships among them
- Understands the basic concept of rate as a measure
- Uses proportional reasoning to solve mathematical and real-world problems

Overview

Students use unit rates to calculate the best deal on different items.

Problem-Solving Strategies

- Organize information in a picture, list, table, graph, or diagram
- Use logical reasoning

Materials

- *Best Buys* (page 37; bestbuys.pdf)
- two containers of the same food item of different sizes, with a price tag to show their costs
- *Student Response Form* (page 132; studentresponse.pdf) *(optional)*

Activate

1. Display two containers of food. Ask students how they can determine which of the two products is a better buy. Record students' suggestions on the board.

2. Ask *If a 12-ounce box of cereal costs $2.99, and a 16-ounce box costs $4.19, which is the better purchase?* (12-ounce box) Have students complete the problem in small groups. Ask volunteers from each group to share their strategies and solutions.

3. If students do not articulate the advantage of finding the unit cost, suggest it, and review with students how to find the unit cost by dividing the total cost by the number of units.

Solve

1. Distribute copies of *Best Buys* to students. Have students work alone, in pairs, or in small groups.

2. As students solve the problems, ask clarifying questions, such as *How else might you think about the problem? How do unit fractions help?*

3. Invite several volunteers to share their solutions and solution paths.

Debrief

1. How do unit rates help you make decisions?

2. What other information might you consider when making a purchase?

Differentiate ⚪

Many students will benefit from making a table beginning with the cost for a unit and increasing the number, weight, or capacity with corresponding price increments.

Jack is helping the student council buy supplies for the school dance. A 32 oz. bag of Crunchy Chips costs $2.99. Jumbo Chips cost $3.99 for 64 oz. Would it be a better buy if Jack bought Crunchy Chips or Jumbo Chips? Justify your answer by using a rate table.

Tien-Li is shopping for a dirt bike. She is trying to decide between Bike A that travels 856 miles on a full tank of fuel and Bike B that travels 915 miles on a full tank of fuel. Bike A has a 12-gallon fuel tank and Bike B has a 13-gallon fuel tank. Which dirt bike is the better buy to save on fuel costs? Justify your answer by using a rate table.

Nick is purchasing DVDs to watch during his vacation. He plans to buy one DVD for each of the thirteen days he will be away. The Game Store has special pricing and will sell 13 DVDs for $63.75, 11 DVDs for $49.99, or 1 DVD for $4.99. Which of the following is the best buy?

a. buying 13 DVDs at the individual price ($4.99 each)

b. buying 11 DVDs for $49.99 and the remaining two at $4.99 each

c. buying 13 DVDs as a group for $63.75

Justify your answer.

Paint Colors

Standards

- Uses proportional reasoning to solve mathematical and real-world problems
- Organizes and displays data using tables, graphs, frequency distributions, and plots

Overview

Students create tables of equivalent ratios and represent ratios on the coordinate plane.

Problem-Solving Strategies

- Organize information in a picture, list, table, graph, or diagram
- Use logical reasoning

Materials

- *Paint Colors* (page 39; paintcolors.pdf)
- graph paper (graphpaper.pdf)
- *Student Response Form* (page 132; studentresponse.pdf) *(optional)*

Activate

1. Distribute graph paper and instruct students to draw an *x*-axis and *y*-axis. Have them label the *x*-axis *water*, the *y*-axis *paint*, and number each axis from 0–25.

2. Ask students to plot the ratio 5:7 on the coordinate plane, using the first term as the *x*-coordinate and the second term as the *y*-coordinate.

3. Instruct students to graph as many equivalent ratios as possible, then connect the points to the origin.

4. Ask students to plot the ratio 1:3 on the same coordinate plane, using the first term as the *x*-coordinate and the second term as the *y*-coordinate.

5. Instruct students to graph as many equivalent ratios as possible, then connect the points to the origin.

6. Ask students to interpret what the graph means. Ask questions such as *Which ratio is greater? How do you know?*

Solve

1. Distribute copies of *Paint Colors* to students. Have students work alone, in pairs, or in small groups.

2. As students work, ask clarifying questions, such as *What other points are on the line? How do they relate to the ratio you graphed? What do you think the steepness of the line means?*

3. Invite several volunteers to share their solutions and solution paths.

Debrief

1. What do you know about how the values for the two colors of paint are related?

2. What other information can you gather from the graphical representation?

3. What might happen if you reversed the labels on the *x*-axis and *y*-axis?

Differentiate

Have students create a ratio table for each ratio to make ordered pairs to plot on the coordinate plane.

Tommy and Renee are mixing paint for art class. Tommy decided to use 5 parts green paint to 7 parts white paint while Renee decided to use 7 parts green paint to 12 parts white paint. Complete the tables of equivalent ratios for Tommy and Renee. Graph the ratios.

Tommy

Green Paint	White Paint
5	7
10	
	21
20	

Renee

Green Paint	White Paint
7	12
	24
	36
28	

Marty wants to make a paint mixture that is darker in color than 1 part blue paint to 4 parts white paint, but not as dark as 5 parts blue paint to 6 parts white paint. Name three mixture ratios that come between the two given mixtures. Use a graph to support your solution.

Mrs. Kolberg, the art teacher, is experimenting with various paint mixtures. She begins with a mixture that was 4 parts red paint to 7 parts white paint. If she adds another mixture to the first one that is 2 parts red paint to 2 parts white paint, will the mixture be darker or lighter red? Justify your answer using a graph, diagram, or table.

Mixing It Up

Standards

- Uses proportional reasoning to solve mathematical and real-world problems
- Organizes and displays data using tables, graphs, frequency distributions, and plots

Overview

Students solve problems using equivalent ratios.

Problem-Solving Strategies

- Organize information in a picture, list, table, graph, or diagram
- Work backward

Materials

- *Mixing It Up* (page 41; mixing.pdf)
- *Student Response Form* (page 132; studentresponse.pdf) *(optional)*

Activate

1. Write the following problem on the board: *There are 15 buttons in a jar in the ratio of 2 blue buttons to 3 green buttons. More blue and green buttons were added to the jar in the same ratio. If there are now 40 buttons in the jar, how many blue and green buttons are there? (16 blue buttons and 24 green buttons)*

2. Have students work in pairs to solve the problem.

3. Ask students to share their answers and strategies. Record them on the board.

4. If no students mention a scale factor, introduce it when comparing the solution. *Notice the ratio 2:3 is scaled by a factor of 8 to reach the ratio of 16:24.*

Solve

1. Distribute copies of *Mixing It Up* to students. Have students work alone, in pairs, or in small groups.

2. As students work, ask clarifying questions, such as *How might you use a ratio table to help you find the new ratio? When the ratio changes, does the total also change?*

3. Invite student volunteers to the board to share their solutions. Ask if anyone got a different answer or if anyone did the problem differently.

Debrief

1. What strategies did you use to solve the problem?

2. What information can you gather from a ratio table?

Differentiate ⬤

Have students create a ratio table or diagram to organize information before solving the problem.

In their bead collection, the ratio of Ashlyn's beads to Avery's beads is 3:4. Ashlyn bought another 108 beads. To keep the same ratio, how many beads does Avery need to buy?

Ari and Ben both have a baseball card collection. The ratio of Ari's cards to Ben's cards is 5:3. Ari gave half of his cards to Ben. Now Ben has 60 more baseball cards than Ari. How many baseball cards did Ari give to Ben? Justify your answer using a ratio table or diagram.

Nolan and Cameron shared some marbles in a ratio of 5:7. Cameron gave Nolan 15 marbles. If each boy had the same number of marbles after sharing, how many marbles did Cameron have to start? Justify your answer using a ratio table or a diagram.

Percent Tables

Standards

- Uses proportional reasoning to solve mathematical and real-world problems
- Understands the concepts of ratio, proportion, and percent and the relationships among them

Overview

Students use percents and percent tables to solve problems.

Problem-Solving Strategies

- Organize information in a picture, list, table, graph, or diagram
- Use logical reasoning

Materials

- *Percent Tables* (page 43; percenttables.pdf)
- *Student Response Form* (page 132; studentresponse.pdf) *(optional)*

Activate

1. Ask students how they might find 30% of 120. Give them time to think individually about how to solve the problem.
2. Have students discuss their strategy with a partner.
3. Ask student volunteers to share their strategies. Ask if anyone did it differently or got a different answer.

Solve

1. Distribute copies of *Percent Tables* to students. Have students work alone, in pairs, or in small groups.
2. As students work, ask clarifying questions, such as *Why did you decide to begin with that percent? Can you work backward from 50%? What strategies can you use to mentally compute percents?*
3. Invite student volunteers to the board to share their solutions. Ask if anyone got a different answer or if anyone did the problem differently. Share various responses.

Debrief

1. How did you choose a starting benchmark?
2. What patterns do you see in the table?

Differentiate △

For students who need a greater challenge, ask them to write a set of steps for finding the percent of any number.

Madison is buying a new bicycle. Complete the table below. How much money did Madison save if she received a 25% discount on a bicycle that cost $120?

Discount	Discount Percent
$120	100%
	50%
	25%

Kevin started to complete a table to show how much money he would save if he received a total of 40% off a computer that originally sold for $499. Finish the table for Kevin. How much money did he save? How much money did he pay for the new computer?

Discount	Discount Percent
$499	100%
	50%
	25%
	10%

Emily bought a new video game system that regularly costs $250. She had a coupon for 15% off but had to pay a sales tax of 6%. How much money did Emily spend?

How Many Groups?

Standard

Divides fractions

Overview

Students use common denominators to model division of fractions using an area model.

Problem-Solving Strategies

- Organize information in a picture, list, table, graph, or diagram
- Use logical reasoning

Materials

- *How Many Groups?* (page 45; howmanygroups.pdf)
- *Student Response Form* (page 132; studentresponse.pdf) *(optional)*

Activate

1. Ask students why we divide. Have students share their thinking with a partner. Record their responses on the board.

2. Distribute paper to students and direct them to fold the paper vertically into fourths to create four columns. Tell students to lightly shade in three-fourths. Next, instruct students to fold the paper horizontally into thirds (to create three rows). Ask students to describe how many parts represent the original three-fourths. ($\frac{9}{12}$) How many parts represent one-third? ($\frac{4}{12}$)

3. Ask students to use the area model to solve the following problem: $\frac{3}{4} \div \frac{1}{3}$. ($\frac{9}{4}$)

4. Invite student volunteers to share their work.

5. Write the same division problem from step 3 vertically: $\dfrac{\frac{3}{4}}{\frac{1}{3}}$

6. Ask students what property they should use to get a denominator of one. (*multiplicative inverse*)

Solve

1. Distribute copies of *How Many Groups?* to students. Have students work alone, in pairs, or in small groups.

2. Encourage students to make an area model to help solve the problem.

Debrief

1. How does an area model help you divide fractions?

2. What happened to the denominators when you made an area model?

Differentiate ⬤ ◼ △ ☆

Assign the following exit card task: *Which is greater $\frac{2}{3} \div \frac{3}{4}$ or $\frac{2}{3} \times \frac{3}{4}$? ($\frac{2}{3} \div \frac{3}{4}$) Justify your answer using an area model or picture.*

Susie is sharing her beads with her friends. She has $\frac{3}{4}$ pounds of beads. If she gives $\frac{3}{8}$ pound to each friend, how many friends will receive beads? Justify your answer with a picture or area model.

Logan has $\frac{7}{8}$ quart of ice cream that he is sharing with his friends. If each bowl holds $\frac{1}{9}$ quart, how many people will get ice cream? Draw a picture or area model to support your answer.

Ms. Moran bought $2\frac{1}{2}$ gallons of ice cream to reward her class. She is planning to distribute $\frac{3}{4}$ pint of ice cream to each student. If she has 30 students, will each student get ice cream? If so, how much ice cream will be leftover? If not, how much more ice cream does she need?

Identical Groups

Standard

Uses number theory concepts to solve problems

Overview

Students determine greatest common factors (GCF) to solve word problems.

Problem-Solving Strategies

- Organize information in a picture, list, table, graph, or diagram
- Simplify the problem

Materials

- *Identical Groups* (page 47; groups.pdf)
- graph paper (graphpaper.pdf)
- inch-tiles *(optional)*
- *Student Response Form* (page 132; studentresponse.pdf) *(optional)*

Activate

1. Distribute graph paper to students. Have them draw a 6 × 9 rectangle on the paper.

2. Have students shade the largest square possible. *(6 × 6)*

3. Have students shade the largest square possible from the remaining 3 × 6 rectangle. Ask them to identify the dimensions of that square. *(3 × 3)* Call attention to the fact that the dimensions of the remaining square are 3 × 3. The side length of the largest remaining square is the greatest common factor (GCF) of 6 and 9. *(3)*

4. Have students use the strategy from step 3 to determine the greatest common factor of 3 and 5 *(1)*, 4 and 6 *(2)*, and 12 and 18 *(6)*. Repeat with other numbers, as necessary.

Solve

1. Distribute copies of *Identical Groups* to students. Have students work alone, in pairs, or in small groups.

2. As students work, ask clarifying questions, such as *Why is that appropriate for this problem? What does it mean when only one square remains?*

3. Invite several volunteers to share their solutions and solution paths.

Debrief

1. How else might you solve the problem?

2. How might you break the problem into a simpler problem before solving it?

Differentiate ⬤

Some students may benefit from using a Venn diagram. Have students decompose each number into prime factors, then place them in the appropriate circles.

Identical Groups

Liam and his friend Ben are making beaded bracelets to sell at the school fair. They have 540 black beads and 600 silver beads. If the boys want to use up all the beads, what is the greatest number of identical bracelets they should make? How many of each color bead will they use per bracelet? Use a picture, table, or diagram to solve.

Identical Groups

Dixon is making piles of coins. He has 16 quarters, 32 dimes, and 40 nickels. If he uses all the coins, what is the greatest number of identical piles he can make? How many of each coin is in each pile? Justify your answer with a picture, table, or diagram.

Identical Groups

Kaya and Emma are making floral baskets for May Day. They have 72 roses, 54 tulips, and 36 carnations. If they use all the flowers to make identical baskets, what is the greatest number of baskets they can make? How many of each flower will each basket have? Justify your answer with a picture, table, or diagram.

Are We in Sync?

Standard

Uses number theory concepts to solve problems

Overview

Students determine least common multiples (LCM) to solve word problems.

Problem-Solving Strategies

- Organize information in a picture, list, table, graph, or diagram
- Simplify the problem

Materials

- *Are We in Sync?* (page 49; sync.pdf)
- *Student Response Form* (page 132; studentresponse.pdf) *(optional)*

Activate

1. Have students count to 30 by 2s. Record the numbers on the board.

2. Have students count to 32 by 4s. Record the numbers on the board.

3. Ask students to identify the common numbers between the lists of numbers.

4. Draw a Venn diagram to show the multiples of 2 and 4 with the common multiples in the intersection. Ask students to identify the least common multiple.

Solve

1. Distribute copies of *Are We in Sync?* to students. Have students work alone, in pairs, or in small groups.

2. As students work, ask clarifying questions, such as *Why is finding the least common multiple appropriate for this problem? What tool could you use to help you solve this problem?*

3. Invite several volunteers to share their solutions and solution paths.

Debrief

1. How might you use a Venn diagram to explain your solution?

2. How might you break the problem into a simpler problem before solving it?

Differentiate △

Ask students to write an equation they can use to find the least common multiple in relation to a product and the greatest common factor ($LCM = \frac{(a \times b)}{GCF}$)

Starting this month, Scott has Judo lessons every second day. His brother Jaden has hockey every third day. Their sister needs to be at baton practice every sixth day. After the first day of the month, what is the next day when all the children are free?

January

Sun.	Mon.	Tues.	Wed.	Thurs.	Fri.	Sat.
	1	2	3	4	5	6
7	8	9	10	11	12	13
14	15	16	17	18	19	20
21	22	23	24	25	26	27
28	29	30	31			

Sam timed the blinking lights at two locations near his school. One light blinks every 4 seconds, and the second light blinks every 6 seconds.

- How many times do the two lights blink at the same time in 60 seconds?
- How many times in five minutes?

Jenny noticed the films at the Indian History Museum were different lengths and run continuously. All four films begin at 8:30 AM. The length of each film is listed on the museum bulletin board.

Film Title	Length
Introduction to the Exhibits	3 minutes
Trail of Tears	24 minutes
Custer's Last Stand	18 minutes
The Battle of Little Big Horn	12 minutes

- What is the next time all four movies will start at the same time?
- Between 8:30 AM–5:00 PM, how many times during the day will all four movies start at the same time?

Are We In Sync?

Factors or Multiples?

Standard

Uses number theory concepts to solve problems

Overview

Students find greatest common factors and least common multiples to solve applied problems.

Problem-Solving Strategies

- Organize information in a picture, list, table, graph, or diagram
- Simplify the problem

Materials

- *Factors or Multiples?* (page 51; factorsmultiples.pdf)
- *Which One Am I?* (whichone.pdf)
- *Student Response Form* (page 132; studentresponse.pdf) *(optional)*

Activate

1. Display *Which One Am I?* Ask students how they might determine whether the problem is asking for a greatest common factor or a least common multiple.

2. Ask students to discuss the strategies they might use to solve the problems. Record the strategies on the board.

3. Have students complete the problems. Invite volunteers from several groups to share their solution strategies and responses.

Solve

1. Distribute copies of *Factors or Multiples?* to students. Have students work alone, in pairs, or in small groups.

2. For students who are struggling, suggest they examine the strategies you listed on the board.

3. Invite students to share their solutions and solution paths.

Debrief

1. What strategies did you use to solve the problem?

2. How might you break the problem into a simpler problem before solving it?

Differentiate ○ □ △ ☆

Assign an exit card task, such as *Explain what the intersection of two circles in a Venn diagram means.*

Factors or Multiples?

You are sending party invitations to your classmates. You want to buy the same number of invitations as stamps. Stamps come in packages of 25. Invitations come in packages of 40. What is the fewest number of packages of stamps and invitations you can buy to make sure you have one stamp for each card? Justify your answer using a diagram or picture.

Factors or Multiples?

The large roller coaster takes 7 minutes from start to finish while the small roller coaster only takes 5 minutes from start to finish. If both roller coasters start at 9:00 AM, when is the next time they will start at the same time? How many times will the rides start at the same time over the next three hours? Make a table or organized list to justify your answer.

Factors or Multiples?

Seamus and Aiden are filling backpacks for the local elementary school. They have 24 boxes of markers, 56 coloring books, and 72 packages of modeling clay. What is the greatest number of backpacks they can fill if the markers, books, and clay are equally distributed among the backpacks? Justify your answer with a diagram, picture, or organized list. Be sure to explain your thinking.

What's My Value?

Standard

Understands the relationships among equivalent number representations

Overview

Students convert fractions to decimals and percents using multiple representations.

Problem-Solving Strategy

Organize information in a picture, list, table, graph, or diagram

Materials

- *What's My Value?* (page 53; value.pdf)
- 10×10 grids
- fraction, decimal, percent graphic organizer (*optional*)
- *Student Response Form* (page 132; studentresponse.pdf) (*optional*)

Activate

1. Ask students how they would write the fraction $\frac{2}{5}$ as a percent. *(40%)*

2. Write the following fractions on the board: $\frac{7}{10}, \frac{3}{5}, \frac{1}{4}, \frac{5}{8}, \frac{2}{3}, \frac{5}{6}$. Ask students what strategies they might use to convert these fractions to percents without using a pencil or paper.

3. Ask students to share their strategies. Record each strategy on the board.

4. Have students find the percent equivalents using one or more of the strategies.

5. Have students share their percent equivalents.

Solve

1. Distribute copies of *What's My Value?* to students. Have students work alone, in pairs, or in small groups.

2. Observe students as they work and note those who model the representations using a 10×10 grid and those who use "friendly" numbers to compute mentally or with paper and pencil.

3. Invite several students to share their solutions and solution paths.

Debrief

1. How might you use your strategies to estimate the equivalent percent when the denominators are not "friendly"?

2. How might you use your strategies for finding a percent given a fraction, and to find a fraction given a percent?

Differentiate ●■▲☆

Assign an exit card task, such as *Represent the fraction $\frac{3}{8}$ as a percent and as a decimal.*

Jessa wants to use grids to show her sister what corresponding percents look like when she knows a fractional amount. Shade in the grids to show how Jessa could represent each fraction as a percent.

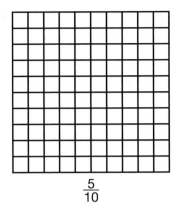
$$\frac{5}{10}$$

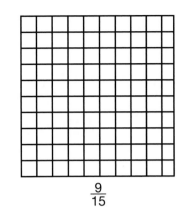
$$\frac{9}{15}$$

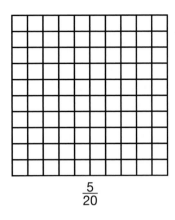
$$\frac{5}{20}$$

Anthony told Jane that he likes to use a bar model to represent fractions as percents. Shade in the bars to show how Anthony could represent the fractions as percents.

$\frac{4}{12}$ [ ] 100%

$\frac{3}{8}$ [ ] 100%

$\frac{5}{6}$ [ ] 100%

William wants to show the fractions $\frac{4}{5}$ and $\frac{6}{8}$ on a grid and in a rectangular bar. How might he display each fraction?

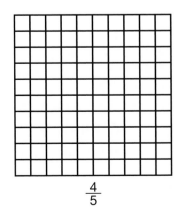
$$\frac{4}{5}$$

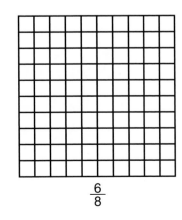
$$\frac{6}{8}$$

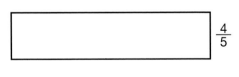
$$\frac{4}{5}$$

$$\frac{6}{8}$$

Greater or Less Than Zero?

Standard

Understands the role of positive and negative integers in the number system

Overview

Students locate positive and negative integers on a number line.

Problem-Solving Strategies

- Act it out or use manipulatives
- Organize information in a picture, list, table, graph, or diagram

Materials

- *Greater or Less Than Zero?* (page 55; greaterless.pdf)
- masking tape
- algebra tiles (*optional*)
- *Student Response Form* (page 132; studentresponse.pdf) (*optional*)

Activate

1. Ask students what they know about negative integers.

2. Use tape to create a number line on the classroom floor.

3. Have students draw a number line on their paper. Mark 0 and positive and negative integers on the number line.

4. Ask a student volunteer to stand on 0 on the number line, then move +5 spaces. Have students do the same on their individual number lines.

5. Ask the volunteer to move –8 spaces. Ask the volunteer to identify his or her location. (–3) Ask if anyone got a different answer or did it differently.

Solve

1. Distribute copies of *Greater or Less Than Zero?* to students. Have students work alone, in pairs, or in small groups.

2. Have students record their work on the number lines.

Debrief

1. What does the sign in front of an integer tell you?

2. How do you determine the direction in which to move?

3. What does a negative integer represent? What does a positive integer represent?

Differentiate ◯ ☆

Provide students who struggle with integers additional practice in moving on a life-sized number line.

At the amusement park, Melia went on a submarine ride. The guide announced that they were going to a depth of 30 feet and would see a lot of fish. How would you describe the submarine's depth? Represent the change in location on the number line below.

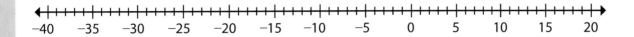

When Grace read the thermometer this morning, it was 32°F. When she read it just before bedtime the temperature was 17°F. What was the change in temperature from morning until bedtime? Represent the change in the temperature on the number line.

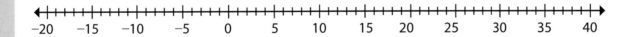

In the mornings, Roger walks 6 blocks east (right) from his house to school. Every Wednesday after school, he walks 13 blocks west (left) to go to soccer practice. How many blocks away from his house is the soccer field? Show your thinking on the number line.

Roger's house

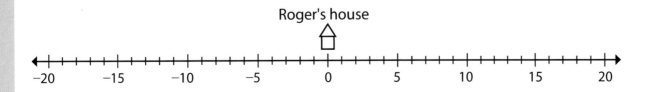

Integer Values

Standards

- Understands the relationships among equivalent number representations
- Understands the characteristics and properties of the set of rational numbers and its subsets

Overview

Students use vocabulary and clues to name numbers or identify the magnitude of numbers.

Problem-Solving Strategies

- Act it out or use manipulatives
- Organize information in a picture, list, table, graph, or diagram
- Use logical reasoning

Materials

- *Integer Values* (page 57; integervalues.pdf)
- decks of playing cards, one per pair
- *Student Response Form* (page 132; studentresponse.pdf) *(optional)*

Activate

1. Write the following pairs of values on the board: –8 ◯ –3; –2 ◯ –11; 12 ◯ 4; –6 ◯ 6; 0 ◯ –5; –10 ◯ –15. Ask students to insert the symbols $<$, $>$, or $=$ between each integer in the pair to make the equation true.

2. Invite students to share their solutions.

3. Distribute a deck of playing cards to pairs of students. Tell them that red cards are negative and black cards are positive. Have students split the deck so each partner receives 26 cards. Each student turns over one card. They compare the value, and the student with the greater value wins the cards.

Solve

1. Distribute copies of *Integer Values* to students. Have students work alone, in pairs, or in small groups.

2. When students complete their problem, have them compare their solutions with a partner.

Debrief

1. What strategies can you use to compare positive and negative integers?

2. If you have multiple negative integers, how can you determine which is greater?

Differentiate ◯

Some students benefit from walking on a number line to gain a deeper understanding of how the values to the left of zero on a number line decrease as you move farther to the left, while the values on the right increase as you move farther to the right.

Sarah and Tim were playing a number game. Sarah drew the cards below. Match Sarah's numbers with the letter on the number line. Which number has the greatest value? Which number has the least value?

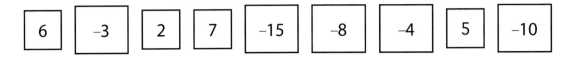

Damien is playing a computer game. He must order the numbers below from smallest to largest, then drag them to the correct location on a number line. Write the numbers below in order from least to greatest.

−4	−21	−11	−35	−18	−1	−3	−13	−9

Olivia is playing a pinball game at an arcade. For every galactic starship she makes contact with, she wins points. If she makes contact with a meteor, she loses points. If she ends up with a positive score when the game ends, she wins a free game. The sequence of points she won and lost during a game included +150, −250, −350, +500, −100, +400, −200, +250, −400, −500, and +300. Did she win a free game?

Opposites Attract

Standards

- Understands the relationships among equivalent number representations
- Understands the characteristics and properties of the set of rational numbers and its subsets
- Understands the role of positive and negative integers in the number system

Overview

Students identify where opposite numbers are located on the number line.

Problem-Solving Strategies

- Act it out or use manipulatives
- Organize information in a picture, list, table, graph, or diagram

Materials

- *Opposites Attract* (page 59; opposites.pdf)
- strips of grid paper
- *Student Response Form* (page 132; studentresponse.pdf) *(optional)*

Activate

1. Distribute a long strip of grid paper to each student. The strip should have at least 30 boxes.

2. Have students fold the grid strip in half. Tell them to label the box in which the fold lies as 0 and number the boxes to the right with consecutive positive integers and to the left with consecutive negative integers.

3. Demonstrate how to fold the strip so that the opposites match up.

4. Ask *What integer is the opposite of –9?* (9) *What integer is the opposite of 4?* (–4) *What integer is the same distance from 0 as –7?* (7)

Solve

1. Distribute copies of *Opposites Attract* to students. Have students work alone, in pairs, or in small groups.

2. Have students record their work on the number lines.

Debrief

1. Why is –10 the opposite of 10?

2. How can you tell which number has the greater value, –5 or –15? 5 or 15?

Differentiate ☆ ◯

Some students struggle with the language of *the opposite of* a number so it is helpful to say it, write it, and notate it as $-(-a) = a$ *(when a > 0)*.

Brent stood on −17 on the class number line. Where would he be standing on the number line if he walked the opposite of −17? Show your answer using the number line.

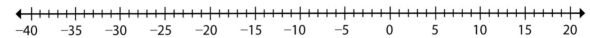

Gloria was at the Mini Mart 18 blocks from her school. She called her dad for a ride home and told him where she was. Where on the number line should her dad look for Gloria? Justify your answer using the number line and a number sentence.

School

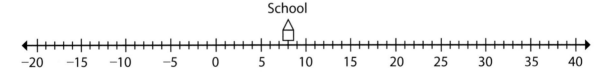

Jorge found a message that showed a cryptic symbolic sentence: $0 + |-13| + -5 = ?$ How might Jorge represent that scenario on a number line?

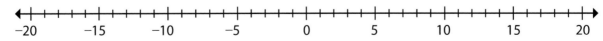

Coordinate Graphing

Standard

Uses the rectangular coordinate system to model and to solve problems

Overview

Students demonstrate an understanding of the coordinate plane, the locations of the four quadrants, the signs of ordered pairs located within each quadrant, and the concept of intercepts.

Problem-Solving Strategies

- Act it out or use manipulatives
- Organize information in a picture, list, table, graph, or diagram

Materials

- *Coordinate Graphing* (page 61; coordinategraphing.pdf)
- *Coordinate Plane* (coordinateplane.pdf)
- *Student Response Form* (page 132; studentresponse.pdf) *(optional)*

Activate

1. Distribute copies of *Coordinate Plane* to students and project it for the class to see.

2. Guide students in labeling the following terms on their individual coordinate grids: *Quadrant I, Quadrant II, Quadrant III, Quadrant IV, origin, x-axis,* and *y-axis.*

3. Invite several volunteers to demonstrate where various points should be plotted. Be sure to include intercepts, such as (0, −3) and (7, 0).

Solve

1. Distribute copies of *Coordinate Graphing* to students. Have students work alone.

2. As students are working, ask them if they can tell by looking at the ordered pairs in which quadrant the points will lie.

Debrief

1. How are the points in Quadrants I and II related? What about III and IV?

2. How do you know which number in the ordered pair to look at first?

Differentiate △

Challenge students to experiment with graphing different orientations of geometric figures in related quadrants. For example, ask *How might a rectangle graphed in Quadrant II look in Quadrant IV if the rectangle has the same dimensions and is positioned the same distance away from each axis?*

Cedric is taking a city tour. The tour starts at the Public Commons, which is located at (–2, –4) and begins by walking three blocks west (left). After pausing to visit a monument, the group continues to walk seven blocks north (up), where they visit a historical cemetery. Then, they walk six blocks to the east (right) where they stop at a historical café for lunch.

How many blocks has the group walked when they stop to eat?

Represent the city tour on a coordinate plane.

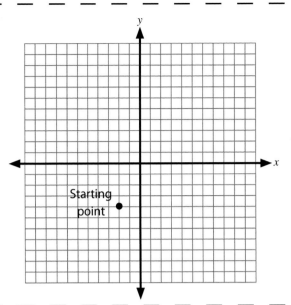

The expression below represents the route Emma took on her weekend bicycle trip. When she traveled west or south, she marked the distances as negative miles. When she went east or north, she recorded the distances as positive miles. She never retraced the same path. She stayed on roadways (grid lines) the whole way.

$$-3 + 5 + 4 - 6 - 3$$

If each grid mark is one mile, how many miles did Emma ride? Trace the route you think she took.

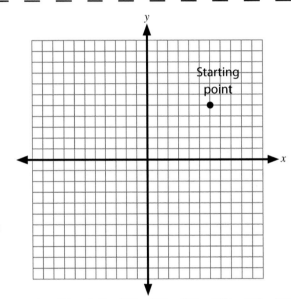

Abby is running errands around town. She starts at the origin and travels north, south, east, and west. If she traveled a total of 62 blocks (on grid lines) and made 5 stops, where might she have stopped? Plot your points on the coordinate plane.

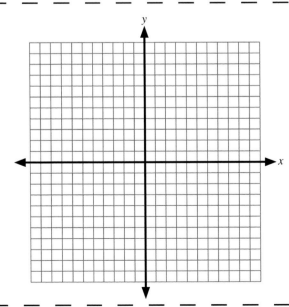

Computing with Integers

Standard

Understands the role of positive and negative integers in the number system

Overview

Students demonstrate an understanding of how to model and add integers using tiles.

Problem-Solving Strategies

- Act it out or use manipulatives
- Organize information in a picture, list, table, graph, or diagram

Materials

- *Computing with Integers* (page 63; computingintegers.pdf)
- *Algebra Mat* (algebramat.pdf)
- algebra tiles or centimeter cubes
- *Student Response Form* (page 132; studentresponse.pdf) *(optional)*

Activate

1. Distribute algebra mats and algebra tiles or centimeter cubes to pairs of students.

2. Direct students to model the expression $-4 + 2$ on the algebra mat.

3. Invite a student to model the expression on the board.

4. Model how to simplify the expression by removing zero pairs (opposites). Record the sum as $-4 + 2 = -2$.

5. Direct students to model the expression $5 + -7$. Invite a student to model the expression on the board.

   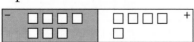

6. Invite a second student to model the sum and record the addition numerically on the board. $5 + -7 = -2$.

   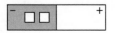

Solve

1. Distribute copies of *Computing with Integers* to students. Have students work alone, in pairs, or in small groups.

2. Invite student volunteers to share their solutions.

Debrief

1. What generalizations can you make about adding integers?

2. Which model for computing with integers works best for you—the number line or the tile model? Why?

Differentiate ○ ▢ △ ☆

As an exit card task, have students list three things they know about working with integers and one question they still have about integers.

Math is Sunhee's favorite subject. She likes modeling expressions on the algebra mat. What expression is on her algebra mat?

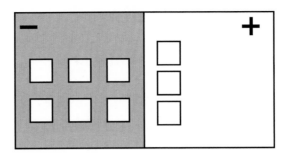

Cameron's teacher asked him to write a mathematical expression for the following representation. What expression might Cameron write?

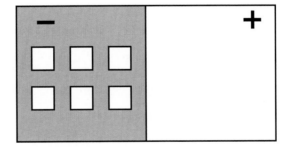

Spencer earned $17 walking his neighbor's dog. He spent most of his money buying baseball cards.

- Write a mathematical equation to show how much money he spent and how much money he had leftover.
- Model your equation with algebra tiles.

What's My Number?

Standard

Understands the characteristics and properties of the set of rational numbers and its subsets

Overview

Students use vocabulary and clues to name numbers with decimal values.

Problem-Solving Strategies

- Guess and check
- Use logical reasoning

Materials

- *What's My Number?* (page 65; whatnumber.pdf)
- *Student Response Form* (page 132; studentresponse.pdf) *(optional)*

Activate

1. Have students say the number 875.9248.

2. Have students identify the place value of each numeral in the number 875.9248.

3. Have students identify the value of each numeral in the number 875.9248.

Solve

1. Distribute copies of *What's My Number?* to students. Have students work alone, in pairs, or in small groups.

2. As students work, listen to the language they use. Model the correct way of reading decimals.

3. Have students draw a place value table so they can model the numbers.

Debrief

1. What did you discover about the difference between naming the place value of numerals to the left and right of the decimal point?

2. How does the value of tenths differ from tens? Hundredths differ from hundreds? Thousandths differ from thousands? What do you think the difference is between millionths and millions?

Differentiate ⬤ ☆

Many students need multiple opportunities to practice naming the place values to the right of the decimal point. It is often helpful to have English language learners make their own place-value charts. Labeling each column repeatedly helps them internalize the order of the place value.

#50778—50 Leveled Math Problems, Level 6

I have 6 hundreds and 24 thousandths. What's my number?

I have 6 thousandths, 9 hundredths, and 4 ones. What's my number?

I have 9 ones and 14 thousandths. What's my number?

I have 6 thousands and 27 thousandths. What's my number?

I have 8 ten-thousands and 12 hundredths. What's my number?

I have 7 tens, 3 hundredths, and 5 ten-thousandths. What's my number?

I have 5 tens, 92 hundredths, and 6 ten-thousandths. What's my number?

I have 429 ones and 86,454 hundred-thousandths. What's my number?

Write your own problem. Share it with a classmate.

Garden Areas

Standards

- Understands the characteristics of a set of rational numbers and its subsets
- Understands formulas for finding measures

Overview

Students use various algorithms to solve problems involving decimals. They compare procedures to determine how decimal values are the same and how they differ, and identify the properties used in determining the solutions.

Problem-Solving Strategies

- Find information in a picture, table, graph, or diagram
- Guess and check or make an estimate
- Organize information in a picture, list, table, graph, or diagram

Materials

- *Garden Areas* (page 67; garden.pdf)
- graph paper (graphpaper.pdf) *(optional)*
- *Student Response Form* (page 132; studentresponse.pdf) *(optional)*

Activate

1. Discuss with students strategies for multiplying 1.3×5. Point out that multiplying 1.3 by 10 for a product *(13)*, then dividing 13 by 2 *(6.5)* is one way to solve this problem.

2. Write 1.3×0.5 on the board. Have students find the answer in two different ways.

3. Have students share their solutions and solution paths.

4. Have students discuss with a partner why the decimal point moves in multiplication but not in addition.

Solve

1. Distribute copies of *Garden Areas* to students. Have students work alone, in pairs, or in small groups.

2. As students work on the problems, listen to the language they use. Model correct vocabulary.

Debrief

1. What did you discover about the relationship between dividing and multiplying decimals?

2. How else might you multiply decimals?

Differentiate ⬤

Many students will need repeated practice developing conceptual understanding of multiplying decimals. Provide many opportunities for students to model decimal multiplication on graph paper.

Shannon is planting a garden. What is the area of her garden?

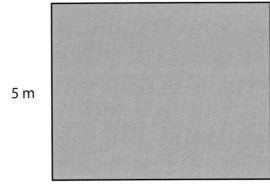

5 m

7.2 m

Riley is trying to find the length of her garden, but she cannot find her measuring tape. She knows the area is 19.32 m². If the width of her garden is 5.6 m, what is the length of her garden?

Rhonda's garden has an area of 19.32 m² and a width of 3.45 m. Mark is trying to explain to Rhonda how to find the length of her garden without dividing. He wrote the following equations:

1 m × 3.45 m = 3.45 m²

10 m × 3.45 m = 34.5 m²

5 m × 3.45 m = 17.25 m²

0.5 m × 3.45 m = 1.725 m²

0.1 m × 3.45 m = 0.345 m²

17.25 m² + 1.725 m² + 0.345 m² = 19.32 m², so the length is 5 m + 0.5 m + 0.1 m = 5.6 m.

Explain how Mark found the area and the length of the garden.

Methods of Multiplying

Standards

- Multiplies integers
- Understands how different algorithms work for arithmetic computations and operations

Overview

Students use multiple procedures for computing to see that they all are valid and produce the same answer when done correctly.

Problem-Solving Strategies

- Count, compute, or write an equation
- Work backward

Materials

- *Methods of Multiplying* (page 69; multiplying.pdf)
- graph paper (graphpaper.pdf) *(optional)*
- *Student Response Form* (page 132; studentresponse.pdf) *(optional)*

Activate

1. Write 34 × 26 on the board. Have students compute the answer *(884)* using two different methods (e.g., traditional algorithm, area model, and expanded notation).

2. Ask students to share their strategies.

Solve

1. Distribute copies of *Methods of Multiplying* to students. Have students work alone, in pairs, or in small groups.

2. Listen to the language the students are using. Restate what they say using place value terminology when necessary. Encourage students to place the 0 in the units place if you see them making an *x* there.

3. Ask students to share their answers. Ask if anyone got a different answer or if anyone did it differently.

Debrief

1. Which method of multiplying makes the most sense to you?

2. If you did not have a calculator handy, which method of multiplying would you use? Why?

Differentiate ◓

For students who have difficulty with organization, you might provide graph paper on which they can place one number in one cell on the paper and line up the digits in the correct place value.

Methods of Multiplying

Nichola bought 23 packages of stickers to use on her art project. Each package of stickers holds 98 stickers. How many stickers does she have? Use two different ways of multiplying.

Methods of Multiplying

Noah bought a total of 54 packages of sports cards. Each of the packages contained 134 cards. How many cards did Noah buy? Use two different ways of multiplying.

Methods of Multiplying

Isaiah noticed that his brother bought 52 bags of marbles. Each bag holds 48 marbles. He bet his brother that he could quickly multiply the two values in his head in under a minute without using paper and pencil. How might Isaiah compute the correct answer quickly? What product should Isaiah get?

Bank On It

Standard

Adds, subtracts, multiplies, and divides integers, and rational numbers

Overview

Students add, subtract, and multiply money.

Problem-Solving Strategies

- Act it out or use manipulatives
- Organize information in a picture, list, table, graph, or diagram

Materials

- *Bank On It* (page 71; bankonit.pdf)
- *Student Response Form* (page 132; studentresponse.pdf) *(optional)*

Activate

1. Write the following problem on the board: *You get a weekly allowance of $5.00. There is a concert you really want to attend. The concert tickets cost $67.50. For how many weeks will you need to save your allowance in order to buy a ticket?* Have students write and solve an equation that represents how long it will take to save enough money to buy the ticket. *(5x = 67.50; It will take 14 weeks to save the money needed.)*

2. Ask a student to share a solution. Ask if anyone else did it differently.

Solve

1. Distribute copies of *Bank On It* to students. Have students work alone, in pairs, or in small groups.

2. As students work on the problems, ask them to explain what absolute value means. Ask students how it is possible to owe money and still give an answer that is positive.

3. Have students share their solutions. Ask if anyone got a different answer or if they did it differently.

Debrief

1. What do you know about absolute value?

2. How might you justify the reasonableness of your answers?

Differentiate ◯□△☆

To help students who may need a more kinesthetic experience, have them model on a class-sized number line what it looks like to owe money versus having a positive balance in a bank account.

Nicky is tracking his deposits and withdrawals. How much money does Nicky have in his bank account?

Deposits	Withdrawals
$21.00	$15.00
$18.00	$5.00
$10.00	$30.00
$21.00	$17.50
$21.00	$23.00

Charlie deposited $17.50 into his school bank account. He needs to buy notebooks, pens, and pencils. He buys 3 notebooks at a cost of $2 each, 3 pens at a cost of $1.50 each, and a 12-pack of pencils for $2. Does he have enough money? If so, how much does he have left? If not, how much money does he need? Support your answer with an equation.

Pedro needs to determine whether the bank is in the black, which means it is not owed any money, or in the red, which means it has lent more money than it has in its account. He knows that Alisha owes $32.20, Jorge owes $22, Taylor has $15 in her account, Alex has $4.50 in his account, Gunner has $12.50, Kai owes $7, Carter owes $1.25, and Lily has $2.50. Based on this information, is the bank running in the red or the black? Support your answer with an equation.

Exponentials

Standard

Understands the characteristics and uses of exponents and scientific notation

Overview

Students represent multiplication expressions with exponential notation.

Problem-Solving Strategies

- Simplify the problem
- Use logical reasoning

Materials

- *Exponentials* (page 73; exponentials.pdf)
- *Exponents* (exponents.pdf)
- tiles and centimeter cubes
- *Student Response Form* (page 132; studentresponse.pdf) *(optional)*

Activate

1. Display *Exponents* for students. Ask them what they know about the expressions. Have students investigate the questions and experiment with writing the expressions using exponents and/or factors.

2. Introduce the terms *base* and *exponent*. Have students identify the bases, explain how the base is used, and explain the role of the exponent.

3. Ask students to share their strategies and solutions.

Solve

1. Distribute copies of *Exponentials* to students. Have students work alone, in pairs, or in small groups.

2. As students are working, ask them to explain how they know which number is the base and which is the exponent.

3. Have students tell you whether a base raised to a zero power is equal to zero. *(No)* Ask them to explain their reasoning.

4. Ask students to share their solutions and solution paths.

Debrief

1. How did you identify the base in the expressions?

2. How did you determine what the exponent would be?

3. How does using exponents in an expression help you?

Differentiate ⬤

Have students model computations using centimeter cubes. They may use unit cubes for numerical representations and rectangular prisms for the variable representations.

Kaleb was absent when his class studied exponents. Help him write the following expressions using exponents.

$8 \times 8 \times 8 \times 8 \times 8 =$ _____

$7 \times 7 \times 7 \times 7 \times 7 \times 7 \times 7 \times 7 \times 7 \times 7 \times 7 \times 7 =$ _____

$4 \times 4 \times 4 =$ _____

Melinda got her quiz on exponents back and saw that she got three problems wrong. Write the correct answer for each problem, and explain to Melinda why each of her answers is incorrect.

$3 \times 3 \times 3 \times 3 = 3 \times 4$ _____

$12 \times 12 \times 12 \times 12 \times 12 \times 12 \times 12 \times 12 \times 12 \times 12 \times 12 = (12 \times 12)^{11}$ _____

$b \times b \times b = 3^b$ _____

One section of a sixth-grade math test was writing expressions in factor form, given exponential form, and writing the exponential form, given expanded form. What should the students write?

$d^6 =$ _____

$b \times b \times b \times b =$ _____

$ab \times ab \times ab \times ab \times ab =$ _____

$c \times c \times c \times c \times c \times c \times c \times d \times d \times d \times d \times d \times d \times d =$ _____

Balancing Act

Standard

Knows that an expression is a mathematical statement using numbers and symbols to represent relationships

Overview

Students balance equations using substitution and logical reasoning.

Problem-Solving Strategies

- Guess and check or make an estimate
- Use logical reasoning

Materials

- *Balancing Act* (page 75; balancing.pdf)
- pan balances (*optional*)
- *Student Response Form* (page 132; studentresponse.pdf) (*optional*)

Activate

1. Display the following on the board: $\triangle + 3 = 9$; $\bullet - 12 = 7$. Have students find the value of $\triangle + \bullet$. *(25)*

2. Ask student volunteers to share their strategies and solution. Ask if anyone used a different strategy or got a different solution.

Solve

1. Distribute copies of *Balancing Act* to students. Have students work alone, in pairs, or in small groups.

2. As students work, listen for academic language. Are they talking about "cancelling?" If so, reiterate what they say by using math language, such as *Subtract from both sides to keep the expressions equivalent.*

Debrief

1. Could you solve the problems in more than one way?

2. What strategies did you find most helpful in determining the weights?

Differentiate ⬤

It may be helpful for some students to use pan balances to weigh various quantities of counters. This helps to get a sense of how objects of different weights can be used to find the weight of unknown objects in relation to the weights of known objects.

Joseph's sister put these shapes on a balance scale. The same shapes always have the same value. What does a  equal?

Mark has different fishing weights. After he weighed them he found some weights balanced others. He drew a picture of which weights balanced each other. The same weights always have the same value.

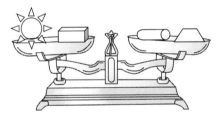
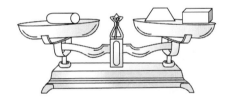

How many will balance ?

Jacinda is looking at two balance scales. The same shapes always have the same value. What does a ◯ equal?

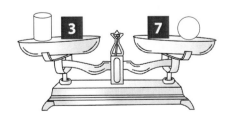
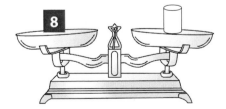

Evaluate Me

Standards

- Knows that an expression is a mathematical statement using numbers and symbols to represent relationships and real-world situations
- Understands basic operations on algebraic expressions

Overview

Students evaluate expressions.

Problem-Solving Strategies

- Guess and check or make an estimate
- Work backward

Materials

- *Evaluate Me* (page 77; evaluateme.pdf)
- whiteboards *(optional)*
- *Student Response Form* (page 132; studentresponse.pdf) *(optional)*

Activate

1. Have students simplify the expression $a + bc + d^2$, if $a = 5$, $b = 3$, $c = 4$, and $d = 6$. (53) Have volunteers share their solutions.

2. Ask if anyone got a different answer or did it differently. Invite those students to share their work.

3. Conduct a round-robin activity. Divide students into groups of three. Assign the number 1, 2, or 3 to each student. Have a number 1 student go to the board as you dictate an expression, such as $ab - c \times d$, if $a = 2$, $b = 10$, $c = 4$, $d = 6$.

- Student number 1 writes the expression, completes the first step, and sits down.

- Student number 2 goes to the board, completes the second step, and sits down.

- Student number 3 goes to the board, completes the next step, and sits down.

- Student number 1 goes to the board, completes the next step, and sits down.

4. Continue the round-robin activity until students find the solution (–4).

Solve

1. Distribute copies of *Evaluate Me* to students. Have students work alone, in pairs, or in small groups.

2. If necessary, have students recall the order of operations.

Debrief

1. How did you know what to substitute into which variables?

2. How did the order of operations help you evaluate the expression?

Differentiate

Give students individual whiteboards to evaluate the expression. Have students write the expression algebraically, then erase the variables and write the value in its place. Remind students they need to use the order of operations to evaluate the expression.

Alex enjoys solving math problems. However, he was confused when he had to evaluate the expression $a - 2^3 + a$, if $a = 15$. Help Alex evaluate the expression. Show your work.

Lisa is evaluating the expression below. She is not sure how to work with the negative signs though. Help her evaluate the expression. Show your work.

$$-a + 3b, \text{ if } a = -3, b = -12$$

Brian is studying for his mathematics test. He knows how to evaluate expressions but thinks he got the problem below incorrect on his review sheet. First, check Brian's answer. If it is incorrect, identify his mistake and show him the correct way. If he is correct, explain why. Show your work.

$$bc + c(3c + 2), \text{ if } b = -3, c = 5$$

$$-3(5) + 5[3(5) + 2]$$

$$-15 + 15(5) + 2$$

$$-15 + 75 + 2$$

$$62$$

Variable Value

Standard

Solves linear equations

Overview

Students solve equations with variables.

Problem-Solving Strategies

- Act it out or use manipulatives
- Count, compute, or write an equation
- Work backward

Materials

- *Variable Value* (page 79; variable.pdf)
- algebra tiles (*optional*)
- balance scales (*optional*)
- equation balance scales (*optional*)
- *Student Response Form* (page 132; studentresponse.pdf) (*optional*)

Activate

1. Review with students that the equal sign means that the two sides are balanced. Display $9 + 2 = \square + 9$. Have students solve the equation. Then, record students' responses on the board.

2. Provide students with balance scales on which to model the problem. Many students may be easily able to answer the problem, but have them explain their answers.

3. Provide students with a series of equations with each equation having a blank in a different location.

4. Have students solve the problems. Have volunteers share their solutions.

Solve

1. Distribute copies of *Variable Value* to students. Have students work alone, in pairs, or in small groups.

2. Ask clarifying questions, such as *How can you make sure that both sides of the equal sign have the same value?*

Debrief

1. How did you decide what the missing value was?

2. What strategies did you use to make sure your answer balances the two sides of the equation?

Differentiate ⬤ ▢ △ ☆

Pair students who are visual or concrete thinkers with a student who is an abstract thinker. Ensure that each student is able to solve the equations concretely as well as symbolically. For students who need a challenge, ask them to identify the properties they used in the solution process. Students should know the commutative, identity, and inverse properties, and how they are used in equations.

Josie is working on solving equations. Help her solve the following equations:

a. $5 + 2 = b + 4$

b. $23 - c = 8 + 6$

c. $17 + 9 = 13 + d$

Justin has to complete the following equations before he goes to lunch. What value does each variable have?

a. $d + 3\frac{1}{2} = 9\frac{1}{2}$

b. $6 + 5 = 11 + b$

c. $7 + e = -7$

Charlie likes algebra. He is solving the equations below. What is the solution for each equation?

a. $m + 7\frac{3}{4} = 13$

b. $24.75 + y = 31$

c. $52 = 18\frac{1}{2} + b$

Systems of Equations

Standard

Solves simple systems of equations

Overview

Students write and solve systems of equations.

Problem-Solving Strategies

- Count, compute, or write an equation
- Use logical reasoning
- Work backward

Materials

- *Systems of Equations* (page 81; systemsequations.pdf)
- pan balances (*optional*)
- *Student Response Form* (page 132; studentresponse.pdf) (*optional*)

Activate

1. Ask students what the term *systems of equations* means. Have them share their responses with the class. If not mentioned, remind students that a system of equations is a set of two or more equations sharing the same unknown values.

2. Ask students to think of an example of a system of equations. Have students share their examples. Record them on the board.

Solve

1. Distribute copies of *Systems of Equations* to students. Have students work alone, in pairs, or in small groups.

2. As students work, ask engaging questions, such as *What patterns do you notice about the figures in each part of the problem? What are some patterns that may help you solve the equations?*

3. Ask clarifying questions, such as *How might you check to ensure your solutions really balance the equations?*

4. Invite students to share their strategies and solutions. Ask if anyone did any of the problems differently or got different answers.

Debrief

1. How did you go about solving the equations?

2. Were you able to solve the equations using logical reasoning, or did you rely on a guess and check approach?

3. How else might you solve these types of problems?

Differentiate ⬤

Kinesthetic learners may benefit from using cutout paper shapes that can be used to model the situation and the solution process. As they work through the modeling, encourage them to record each step they are doing using symbolic notation.

Help Vedant find the value of each shape. The same shapes have the same value.

$\triangle + \bigcirc = 35$

$\triangle + \triangle = 26$

$\square + \bigcirc = 41$

$\triangle = $ _____ $\bigcirc = $ _____ $\square = $ _____

Help Heather find the value of each shape. The same shapes have the same value.

$\square + \bigcirc = 22$

$\triangle + \square = 18$

$\triangle + \bigcirc = 26$

$\triangle = $ _____ $\bigcirc = $ _____ $\square = $ _____

Help Erin find the value of each shape. The same shapes have the same value.
Then, make up your own system of equations and challenge a classmate to solve it.

$\triangle + \bigcirc + \triangle = 42$

$\square + \bigcirc + \triangle = 33$

$\square + \bigcirc + \bigcirc = 30$

$\triangle = $ _____ $\bigcirc = $ _____ $\square = $ _____

Systems of Equations

Heads and Feet

Standard

Solves simple systems of equations

Overview

Students set up and solve equations in which one variable increases while another variable decreases.

Problem-Solving Strategies

- Count, compute, or write an equation
- Organize information in a picture, list, table, graph, or diagram

Materials

- *Heads and Feet* (page 83; headsfeet.pdf)
- images of two- and four-legged animals (*optional*)
- *Student Response Form* (page 132; studentresponse.pdf) (*optional*)

Activate

1. Divide students into groups of three by assigning the number 1, 2, or 3 to each student.

2. Say *There are 7 heads and 22 feet visible in the pasture. Some are cows and some are ducks. How many cows and how many ducks are there?* (*4 cows and 3 ducks*)

3. Student number 1 writes an equation and completes the first step. Student number 2 completes the next step. Student number 3 completes the next step. Complete this rotation until the question has been answered.

Solve

1. Distribute copies of *Heads and Feet* to students. Have students work alone, in pairs, or in small groups.

2. As students work, check to see if they are organizing their data in a table, which is an efficient way of solving these problems.

3. Ask clarifying questions, such as *How might you organize your data to enable you to track the total number of heads that belong to the two-legged animals and those that belong to the four-legged animals? What do you notice about the relationship of the number of two-legged animals to the four-legged animals?*

Debrief

1. What strategies did you find most helpful in figuring out how many of each animal there were?

2. How would you explain the relationship of the number of four-legged animals to two-legged animals?

Differentiate ⬤

Provide a simplified problem for struggling students to work with initially. Display images of two- and four-legged animals as a visual for students who may benefit from them.

At the farm, Marco was looking at the hens and goats to count the number of heads and feet. He counted 40 heads and 114 feet. How many hens and goats did he see? Justify your answer with a table, equation, or picture.

Isabelle works at a zoo. Her job is to take care of the zebras and ostriches that live in neighboring habitats. She counted 46 heads and 154 feet. How many zebras and ostriches did she see? Justify your answer with a table, equation, or picture.

Dakota likes to visit the local ranch to watch the cows and roosters in the pasture. She counted 226 feet and 65 heads. How many cows and roosters did she see? Justify your answer with a table, equation, or picture.

How Much Money?

Standard

Understands the basic concept of a function

Overview

Students use informal methods for setting up and solving equations in which one variable increases while another variable decreases.

Problem-Solving Strategies

- Act it out or use manipulatives
- Guess and check
- Organize information in a picture, list, table, graph, or diagram
- Use logical reasoning

Materials

- *How Much Money?* (page 85; howmuchmoney.pdf)
- play money *(optional)*
- *Student Response Form* (page 132; studentresponse.pdf) *(optional)*

Activate

1. Ask students to determine the greatest number of coins you can have and not be able to make change for a dollar bill. *(99 pennies)*

2. Tell students to think of a strategy they can use to answer the question. Have students share their strategy with a partner. Ask students to share their strategies. List all strategies on the board.

3. Have students use one of the strategies on the board as they work in pairs or small groups to solve the problem.

4. Ask students to share a strategy and a solution. Ask if anyone did it differently, or if anyone got a different answer.

5. Discuss the advantages and disadvantages of each method.

Solve

1. Distribute copies of *How Much Money?* to students. Have students work alone, in pairs, or in small groups.

2. As students are working, ask clarifying questions, such as *How might you organize the data you are collecting? What patterns are you noticing?*

Debrief

1. What strategies did you find most helpful in figuring out how many of each coin you need?

2. How else might you determine the number of coins?

Differentiate ⬤

Provide play money to allow students to model the problem, if necessary.

Tucker has $122.00 in one-, five-, and ten-dollar bills. If he has 15 bills in all, how many of each bill does he have?

Ellie has 13 coins in her wallet. If she has a total of $2.80, what coins might she have?

Shelby and her mom are collecting ticket money at the middle school basketball game. Adult tickets cost $7.00 and student tickets cost $5.00. How many adults and how many students attended the game if 96 people attended and the proceeds for the game totaled $526.00?

My Equation Is

Standard

Solves linear equations

Overview

Students set up and solve equations using variables based on the context of a problem.

Problem-Solving Strategies

- Count, compute, or write an equation
- Organize information in a picture, list, table, graph, or diagram

Materials

- *My Equation Is* (page 87; myequation.pdf)
- graph paper (graphpaper.pdf) *(optional)*
- *Student Response Form* (page 132; studentresponse.pdf) *(optional)*

Activate

1. Say *Myles hiked 37 more miles than his friend Pete. Together, the boys hiked 95 miles. How many miles did Pete hike?* *(29 miles)* Have students write an equation to determine the distance using a variable.

2. Have volunteers share their equations. If any students used different equations, have them share their equations.

3. Discuss the equations students shared. Be sure students define the variable they used.

4. Have students solve the equation.

Solve

1. Distribute copies of *My Equation Is* to students. Have students work alone, in pairs, or in small groups.

2. As students work, be sure they define the variables and substitute the value for the variable in each place the variable appears.

Debrief

1. How did you know where to place the variable in the equation?

2. How did you decide which variable to use for each person in the problem?

Differentiate ⬤

Have students use a bar diagram to model the problem.

My Equations Is

Randi read 28 minutes longer than her brother, Tyler. If they read for 152 minutes total, how long did they read individually? Write and solve an equation.

My Equation Is ▢

Desmond and his sister Penny read for a total of 584 minutes over the course of a month. If Desmond read 122 more minutes than Penny, how many minutes did they read individually? Write and solve an equation.

My Equation Is ◁

Damien and Scott read for a total of $3\frac{1}{2}$ hours over two weeks. If Damien read 124 more minutes than Scott, how many minutes did each read individually? Write and solve an equation.

Arithmetic Sequences

Standard

Understands the properties of arithmetic and geometric sequences

Overview

Students solve linear sequences using recursive or explicit relations.

Problem-Solving Strategies

- Generalize a pattern
- Organize information in a picture, list, table, graph, or diagram

Materials

- *Arithmetic Sequences* (page 89; arithmetic.pdf)
- graph paper (graphpaper.pdf)
- *Student Response Form* (page 132; studentresponse.pdf) *(optional)*

Activate

1. Display an arithmetic sequence on the board. Ask students to think about the strategies they might use to identify a rule for the sequence.

2. Have students add values to the sequence. Record their answers on the board.

3. Distribute graph paper to students. Have students graph the relationship between the number of the term and the value of the term.

4. Ask students to share their graphs. If any students graphed it differently, ask them to share their graphs.

Solve

1. Distribute copies of *Arithmetic Sequences* to students. Have students work alone, in pairs, or in small groups.

2. While students are working, ask clarifying questions, such as *Would it help to find the difference between consecutive terms? How does the number in the sequence relate to its term's position?*

Debrief

1. What strategies did you find most helpful in figuring the remaining values in the sequence?

2. How else might you represent the sequence in order to more clearly determine the rule for the sequence's pattern?

Differentiate ◯ ◻ △ ☆

Students may benefit from graphing the relationship between the term number and the numerical value to help them see that the arithmetic sequences are all linear. Graphing is especially helpful if the values in the sequence are decreasing, which also introduces students to the concept of negative slope.

Arithmetic Sequences

What are the next three terms in each arithmetic sentence? Write the rule for each sequence.

7, 14, 21, 28 , _____, _____, _____

Rule: _____

3, 5, 7, 9, 11, 13, 15, _____, _____, _____

Rule: _____

What are the next three terms in each arithmetic sentence? Write the rule for each sequence.

2, 7, 12, 17, _____, _____, _____

Rule: _____

11, 15, 19, 23, _____, _____, _____

Rule: _____

What are the next three terms in each arithmetic sentence? Write the rule for each sequence.

4, 10, 16, 22, _____, _____, _____

Rule: _____

−1, −4, −7, −10, _____, _____, _____

Rule: _____

Going the Distance

Standards

- Uses proportional reasoning to solve mathematical and real-world problems
- Understands the basic concept of rate as a measure

Overview

Students solve problems using the distance formula.

Problem-Solving Strategies

- Count, compute, or write an equation
- Organize information in a picture, list, table, graph, or diagram

Materials

- *Going the Distance* (page 91; distance.pdf)
- graph paper (graphpaper.pdf)
- *Student Response Form* (page 132; studentresponse.pdf) *(optional)*

Activate

1. Write the following problem on the board: *If I travel to a music concert that is 57 miles from my home, and it takes me 45 minutes to get to the concert, how would I find the rate of speed at which I traveled?* $\left(\frac{57 \text{ miles}}{45 \text{ min.}}\right)$

2. Have students show how they would represent the information in a table, graph, or equation.

3. Have students share their strategies. If any students did it differently, have them share their strategies.

4. Distribute graph paper to students. Then, have students graph the information on the coordinate plane.

5. Have students share their graphs. Ask whether it matters if the number of the term is placed on the *x*-axis or the *y*-axis.

Solve

1. Distribute copies of *Going the Distance* to students. Have students work alone, in pairs, or in small groups.

2. As students work, ask clarifying questions, such as *Does the distance change if your return route is the same?*

Debrief

1. What did you notice about the relationship between the time and the rate?

2. What happened to the rate when the time was changed?

Differentiate ⬤

Provide students with multiple opportunities to solve distance problems in which the missing variable changes.

Logan went with his soccer team to an amusement park. The bus driver drove at an average rate of 60 miles per hour. The team arrived at their destination $2\frac{1}{2}$ hours after leaving. How far did the team travel? Support your answer with a diagram or picture.

Isly went with her friends to a concert. They traveled a distance of 135 miles each way. Make a table to show how long it would take them to arrive at the concert if the car traveled at 40 miles per hour, 50 miles per hour, 60 miles per hour, and 65 miles per hour. Round your answer to the nearest tenth of an hour. Explain what happens to the time it takes compared to the speed the car travels.

Madison and her family drove 165 miles to the beach. Make a table to show the speed her father traveled if it took him $3\frac{1}{2}$ hours, 4 hours 20 minutes, and $2\frac{1}{2}$ hours. Round your answer to the nearest mile per hour. Write an equation to show how fast he was traveling if it took h hours.

Arithmetic and Geometric Sequences

Standard

Understand the properties of arithmetic and geometric sequences

Overview

Students specify terms of geometric sequences using recursive and explicit relations.

Problem-Solving Strategies

- Generalize a pattern
- Organize information in a picture, list, table, graph, or diagram

Materials

- *Arithmetic and Geometric Sequences* (page 93; arithgeoseq.pdf)
- graph paper (graphpaper.pdf)
- *Student Response Form* (page 132; studentresponse.pdf) *(optional)*

Activate

1. Display the sequence 81, 27, 9, 3, 1, $\frac{1}{3}$, $\frac{1}{9}$ on the board. Ask students to think about the strategies they would use to identify a rule for the sequence.

2. Then, ask students to identify the rule for the sequence. *(81 ÷ 3^n)*

3. Have students complete the sequence 4, 16, 64, _____, _____, _____. *(4, 16, 64, 256, 1,024, 4,096)*

4. Ask students to share their sequences. Ask if anyone got a different answer or if anyone did it differently.

5. Distribute graph paper to students. Have students graph the relationship between the number of the term and the value of the term.

6. Ask students to share their graphs and describe the shape.

Solve

1. Distribute copies of *Arithmetic and Geometric Sequences* to students. Have students work alone, in pairs, or in small groups.

2. As students work, ask clarifying questions, such as *Are those differences constant? How might you check to see if they are geometric or arithmetic sequences? How does the number in the sequence relate to its term's position?*

Debrief

1. How did you determine various terms in the sequence?

2. What is the difference between an arithmetic and geometric sequence?

Differentiate ◐ ▢ △ ☆

Students may benefit from graphing the relationship between the term number and the numerical value to help them see that geometric sequences involve multiplication or division.

Arithmetic and Geometric Sequences

Zachary needs to find the 8th and 15th terms in the sequences below. Are the sequences arithmetic or geometric? Find the terms.

3, 7, 11, 15, 19   8th term _____   15th term _____

$\frac{1}{2}$, 1, 2, 4, 8   8th term _____   15th term _____

Arithmetic and Geometric Sequences

Marissa needs to find the 7th and 10th terms in the sequences below. Are the sequences arithmetic or geometric? Find the terms.

$\frac{1}{64}$, $\frac{1}{32}$, $\frac{1}{16}$, $\frac{1}{8}$ 7th term _____   10th term _____

−18, −12, −6   7th term _____   10th term _____

Arithmetic and Geometric Sequences

David needs to complete the sequences below. Are the sequences arithmetic or geometric? Find the terms.

$\frac{2}{9}$, $\frac{2}{3}$, 2, 6, _____, _____, _____

Complete the sequence below. What is the rule for this sequence?

1, 1, 2, 3, 5, _____, _____, _____

Rule: _____

Various and Sundry Patterns

Standard

Understands the properties of arithmetic and geometric sequences

Overview

Students apply arithmetic and geometric sequences to solve problems.

Problem-Solving Strategies

- Organize information in a picture, list, table, graph, or diagram
- Simplify the problem

Materials

- *Various and Sundry Patterns* (page 95; varioussundry.pdf)
- graph paper (graphpaper.pdf)
- *Student Response Form* (page 132; studentresponse.pdf) *(optional)*

Activate

1. Write the sequences 4, 8, 12, 16, 20, and 5, 8, 11, 14, 17 on the board. Ask students to think about the strategies they might use to determine if the sequences are arithmetic or geometric.

2. Ask students to explain the difference between an arithmetic sequence and a geometric sequence.

3. Ask students to share more of each sequence.

4. Distribute graph paper to students. Have students graph the relationship between the number of the term and the value of that term for each sequence.

5. Ask students to share their graphs.

Solve

1. Distribute copies of *Various and Sundry Patterns* to students. Have students work alone, in pairs, or in small groups.

2. As students are working, ask clarifying questions, such as *Are those differences constant? How might you check to see if the sequence is geometric?*

Debrief

1. How does the graph of an arithmetic sequence differ from the graph of a geometric sequence?

2. How might you solve a simpler problem to enable you to find the missing terms?

Differentiate ◯

Have students start with a simpler problem. It is helpful to ensure these students organize their data in an organized list or chart in order to see the patterns that develop.

Emmett and Reese found the following pattern in the book they are reading:

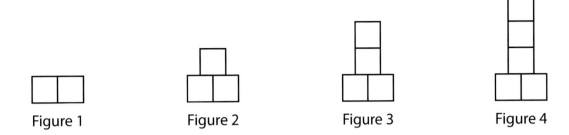

Figure 1   Figure 2   Figure 3   Figure 4

- How many squares will there be in the 5th figure? 10th figure?
- What is the rule?

Ramon made the following pattern using his pattern blocks. His brother Cameron was curious about the pattern. Answer the questions he asked about it.

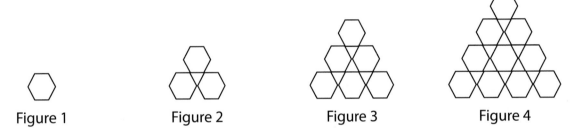

Figure 1   Figure 2   Figure 3   Figure 4

- How many ⬡ will there be in the 7th figure? 10th figure?
- How would you describe this pattern in words?

Jesse made a series of structures using marbles which she glued together. They grew in the following pattern. If the pattern continues to grow, how many marbles will there be in the 7th, 15th, and nth figure? Write an equation to represent the pattern.

Figure 1   Figure 2   Figure 3   Figure 4

Expressly What?

Standard

Understands basic operations on algebraic expressions

Overview

Students recognize and name the various properties of operations.

Problem-Solving Strategies

- Guess and check or make an estimate
- Use logical reasoning

Materials

- *Expressly What?* (page 97; expresslywhat.pdf)
- *Student Response Form* (page 132; studentresponse.pdf) *(optional)*

Activate

1. Pose the following problem to students: *I'm thinking of a number. If you subtract 5, and divide by 4, you get 5.*

2. Have students find the number. *(25)*

3. Ask students to share their solutions.

4. Ask if anyone did it differently or got a different answer. Have those students share their answers. Discuss the solutions and which properties they did or did not use to find the solution.

Solve

1. Distribute copies of *Expressly What?* to students. Have students work alone, in pairs, or in small groups.

2. As students work, ask them if there is more than one equivalence.

3. Ask clarifying questions, such as *What does the distributive property do?*

4. Ask for volunteers to share their solution. Ask if anyone got a different answer or did it differently.

Debrief

1. Why was it important to think about the order of operations?

2. Why did you think it is important to include parentheses?

3. How did you know what operations to use?

4. What were the inverse operations you used?

Differentiate ⬤ ◼ △ ☆

To help students gain experience and practice with the properties of operations, make a set of index card pairs, one with one form of a property, and a second with a different example. Make enough cards that each student in the class receives a card. Once each student has a card, instruct them to find a student who has a card that matches their property. For example, one card may have $33 + 11b$ and a corresponding card may have $11(3 + b)$. Students must explain how they know they made a correct match.

Billy and Tonya are guessing each other's numbers. For Billy's number, he says, "If you add 8 and multiply by 4, you get 56." Tonya replies, "That's easy. 56 − 8 = 48 and $\frac{48}{4}$ = 12. So your number is 12."

Billy says, "No, my number is 6."

How did Billy get 6? Write an expression to support your response.

Eden and Ariel are guessing each other's numbers. For Eden's number, she says, "If you multiply by 2, subtract 8, and multiply by 6, you get 36." Ariel replies, "I can do that. $\frac{36}{2}$ = 18 and $\frac{18}{6}$ = 3 and 3 plus 8 = 11. So your number is 11."

Eden says, "No, my number is 7."

How did Eden get 7? Write an expression to support your response.

Kathryn and Katie are guessing each other's numbers. For Kathryn's number, she says, "If you subtract 10, multiply by 5, and add 4, you get 19." Katie replies, "Oh, I can do that. 19 + 10 = 29 and 29 − 4 = 25. $\frac{25}{5}$ is 5. So your number is 5."

Kathryn says, "No, my number is 13."

How did Kathryn get 13? Write an expression to support your response.

Simplify Me

Standard

Understands basic operations on algebraic expressions

Overview

Students substitute and simplify to evaluate expressions.

Problem-Solving Strategy

Use logical reasoning

Materials

- *Simplify Me* (page 99; simplifyme.pdf)
- *Student Response Form* (page 132; studentresponse.pdf) *(optional)*

Activate

1. Conduct a round-robin activity. Divide the class into groups of three by assigning the number 1, 2, or 3 to each student. Dictate the expression $4b + 3(b + 2)$, if $b = -5$.

2. Student number 1 writes the expression and completes the first step. Student number 2 completes the next step. Student number 3 completes the next step.

3. Continue the round-robin activity until the expression is simplified. *(–29)* Conduct round-robin activities using the same expression for $b = -3$, $b = 6$, $b = 4$, and $b = \frac{1}{2}$. *(–15, 48, 34, 9 $\frac{1}{2}$)*

Solve

1. Distribute copies of *Simplify Me* to students. Have students work alone, in pairs, or in small groups.

2. As students work, observe which students substitute only into the first variable.

3. For struggling students, have them rewrite each expression and replace every variable with a pair of parentheses before they begin the simplifying process.

4. Ask clarifying questions, such as *What does it mean in an expression if a value is given for a specific variable?*

Debrief

1. Did you substitute the value for a variable every time it appeared?

2. What were some of the steps that you used when substituting and simplifying expressions?

Differentiate ⬤

Many students do not realize that they must substitute a given value assigned to a variable each time the variable appears in the expression. To help those students, suggest that they rewrite the expression using parentheses each time they see a variable. Then, in the second step, encourage the students to begin the computation using order of operations.

Marty was asked to simplify the expressions below. What answers should he get? Justify your responses by showing all your steps.

a. $6x - 4 + 7x$

if $x = 3$

b. $19 + 5b + 2b$

if $b = -4$

Zara has to simplify the following expressions but she gets confused when the same variable appears more than once. What should her answers be? Justify your responses by showing all your steps.

a. $-3(2c + 5) + 6c$

if $c = 3$

b. $y + 2(4 + 5y) + 3y$

if $y = -2$

Dina needs to simplify the expressions below. What answers should she get? Justify your responses by showing all your steps.

a. $1 \div d + 8d - 3\frac{1}{2}$

if $d = 4$

b. $5e - 3(4e - 2 + e) + 7$

if $e = \frac{1}{2}$

Simplify Me

Simplify Me

Simplify Me

How Far Did I Go?

Standards

- Uses proportional reasoning to solve mathematical and real-world problems
- Understands the basic concept of rate as a measure

Overview

Students use the distance formula to represent one relationship between dependent and independent variables, and solve equations that are represented.

Problem-Solving Strategies

- Count, compute, or write an equation
- Find information in a picture, list, table, graph, or diagram

Materials

- *How Far Did I Go?* (page 101; howfar.pdf)
- graph paper (graphpaper.pdf)
- *Student Response Form* (page 132; studentresponse.pdf) *(optional)*

Activate

1. Ask students what they know about the equation $d = rt$. Record students' responses on the board.

2. Write $d = 15t$ on the board. Say *A group of students competed in a long-distance bike race and averaged 15 miles per hour each day.* Have students determine how many miles the students traveled after 2 hours, $4\frac{1}{2}$ hours, $8\frac{3}{4}$ hours, and 15 hours. *(30 miles; 67.5 miles; 131.25 miles; 225 miles)*

3. Have students graph the miles on the coordinate plane.

4. Ask students to share their graphs. Ask if anyone graphed it differently.

Solve

1. Distribute copies of *How Far Did I Go?* to students. Have students work alone, in pairs, or in small groups.

2. As students work, ask clarifying questions, such as *How might you ensure that your graph makes sense for the data?*

Debrief

1. How did you use the distance formula?

2. How did you know what values to graph?

Differentiate

Many students benefit from graphing distance, rate, and time relationships but need help in identifying the labels on each axis. Ask them if the rate depends on the time or distance, or if the distance depends on the time and the rate.

Morgan's dad commutes to work. His average rate of speed is 65 mph. How far will he drive in 2 hours? $3\frac{1}{2}$ hours? 5 hours? $6\frac{1}{2}$ hours? Write an equation to justify your responses.

Michaela's mom recorded the total distance and time she travels in a month in the table below. On average, how fast did her mother drive each month? Write an equation to show how far she would have traveled after t hours.

Distance	Time
450 miles	6 hours
337.5 miles	4 hours, 30 minutes
206.25 miles	2 hours, 45 minutes
675 miles	9 hours
150 miles	2 hours

Write a story problem for the following graph. Explain the relationship between the independent and dependent variables.

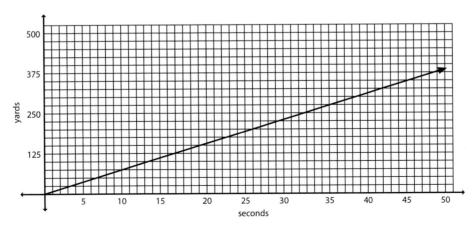

Equivalences

Standard

Understands basic operations on algebraic expressions

Overview

Students substitute and simplify to evaluate expressions.

Problem-Solving Strategies

- Count, compute, or write an equation
- Simplify the problems
- Work backward

Materials

- *Equivalences* (page 103; equivalences.pdf)
- tag board
- tape
- *Student Response Form* (page 132; studentresponse.pdf) *(optional)*

Activate

1. Before the lesson, write $5b$, b, $3b$, 4, $\div$, $+$, $\times$, and $-$ on 8" $\times$ 11" sheets of tag board.

2. Distribute the sheets of tag board to eight students.

3. Have students stand in a line so that an operation comes between a number or variable. Tell them to attach their paper to the board in that order.

4. Assign a numerical value for the value of b, and ask students to simplify the expression on the board by writing an equivalent expression.

5. Have students share their solutions. Ask if anyone got a different answer or did it differently.

6. Repeat the activity by having eight more students arrange themselves in a different order. Have students find the equivalent expression.

Solve

1. Distribute copies of *Equivalences* to students. Have students work alone, in pairs, or in small groups.

2. As students work, ask clarifying questions, such as *How do you know which operations to do first? What does it mean when a number is right next to a variable? How do you know your expression is equivalent?*

Debrief

1. What was the most challenging part of finding equivalent expressions?

2. How did you check your work to ensure computational accuracy?

Differentiate ⬤ ◼ △ ☆

Play "hopscotch" order of operations. Make a hopscotch board on the floor. Students hop on the first squares, but put both feet down at the same time for multiplication and division and addition and subtraction to illustrate the inverses that have the same priority.

Kelly evaluated the expression below. Did she do it correctly? If not, what is her mistake? If so, how do you know? What is the correct answer?

$$8x \div 8 + 7x, \text{ if } x = 4$$

$$8(4) \div 8 + 7(4)$$

$$4 \div 8 + 7(4)$$

$$\tfrac{1}{2} + 28$$

$$28\tfrac{1}{2}$$

Raoul did not study for his math test. When he had to find an equivalent expression for $1 \div c + -3 + 5c$ if $c = -4$, he made a serious mistake. Help Raoul find his error. What is the correct answer?

$$1 \div -4 + -3 + -20$$

$$-4 + -3 + -20$$

$$-7 + -20$$

$$-27$$

Jeremy was asked to write equivalent expressions for the expression below. He asked his teacher how many he should write. What are some of the equivalences he might include?

$$(-b)^2 + 2(-1 + 5b) \div 3b, \text{ if } b = -6$$

Equivalences

Equivalences

Equivalences

Quadrilaterals and Triangles

Standards

- Solves problems involving area of various shapes
- Understands formulas for finding measures

Overview

Students find the area of quadrilaterals and triangles.

Problem-Solving Strategies

- Act it out or use manipulatives
- Organize information in a picture, list, table, graph, or diagram

Materials

- *Quadrilaterals and Triangles* (page 105; quadtri.pdf)
- rulers
- scissors
- graph paper (graphpaper.pdf)
- *Student Response Form* (page 132; studentresponse.pdf) *(optional)*

Activate

1. Ask students to tell you what they know about the relationship between quadrilaterals and triangles.

2. Distribute graph paper, rulers, and scissors to students. Have them cut the paper into four equivalent triangles.

3. Encourage students to work in pairs to share their thinking and strategies. Once students have identified the four triangles, instruct them to prove that the areas are equivalent.

4. Ask student volunteers to share their solutions and strategies they used to prove the four triangles' areas are equivalent. Record each strategy on the board.

Solve

1. Distribute copies of *Quadrilaterals and Triangles* to students. Have students work alone, in pairs, or in small groups.

2. As students work, ask them to predict which area is greatest.

Debrief

1. What strategies did you find helpful when finding the area?

2. Why do you think the units for area are square units? Why is it different from the label used to describe the perimeter?

Differentiate

Have students draw quadrilaterals and triangles on graph paper so they can count the square units that make up the area. Allow them to do this until they are able to articulate a more efficient way of finding the area.

Serena is studying for a math test. She has to find the area of the figures below. What strategies might she use? What is the area for each figure?

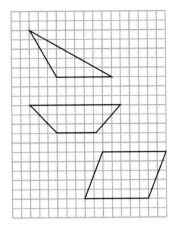

Billy's favorite subject is geometry. He likes the challenge of working with all types of shapes. He is working on finding the areas for the shapes below. Predict which figure has the greatest area. Calculate the area for each figure.

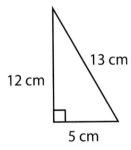

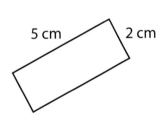

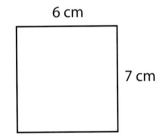

Maria has to find the area of three figures for her math homework. Her teacher told her to predict which of the figures has the greatest area. Which figure do you predict has the greatest area? Find the area of each figure.

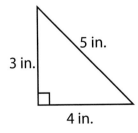

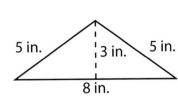

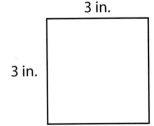

Quadrilaterals and Triangles

Boxy Areas

Standards

- Solves problems involving area of various shapes
- Understands formulas for finding measures

Overview

Students find the area of irregular polygons.

Problem-Solving Strategies

- Act it out or use manipulatives
- Organize information in a picture, list, table, graph, or diagram
- Work backward

Materials

- *Boxy Areas* (page 107; boxyareas.pdf)
- graph paper (graphpaper.pdf)
- *Student Response Form* (page 132; studentresponse.pdf) *(optional)*

Activate

1. Distribute graph paper to students and ask them to trace their hand on it with their fingers spread apart. Have students find the area their hand covers.

2. Ask students to share the solutions and strategies they used to calculate the area. Record and discuss each strategy.

3. If students do not mention making a rectangle to enclose the tracing, calculating the area of the rectangle, and subtracting the squares outside of the hand tracing, point this out as a strategy that may be helpful.

Solve

1. Distribute copies of *Boxy Areas* to students. Have students work alone, in pairs, or in small groups.

2. As students work on the problems, observe the strategies they use.

Debrief

1. What strategies did you find helpful when finding the area of irregular shapes?

2. Why do you think the unit for area is square units? Why is it different from the label used to describe the perimeter?

Differentiate

Guide students in breaking apart irregular polygons into more familiar shapes, such as rectangles or triangles. Help them see that the total area is the sum of the areas of the individual shapes which make up the figure.

Lucia is helping her dad tile an area of the basement. The length of one wall is 9 feet, its opposite side has a length of 13 feet, and the width is 5 feet. What is the area of the space to be tiled? Justify your answer by showing and explaining how your determined the area.

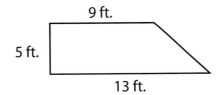

Henry is helping his mom design their summer garden. Henry likes to make designs with the flowers. He needs 12 plants for every square foot of his design. How many plants will he need to fill the T-shape in the garden? Justify your answer by showing and explaining how your determined the area.

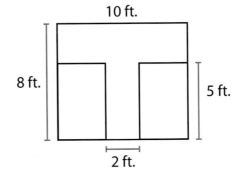

Brett knows the perimeter of the following figure is 108 cm, and the side length of one square is 6 cm. What is the area? Justify your answer.

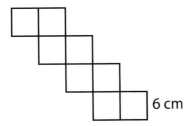

Dot Polygons

Standards

- Understands procedures for basic indirect measurements
- Solves problems involving area of various shapes

Overview

Students find the area of two-dimensional shapes.

Problem-Solving Strategies

- Generalize a pattern
- Organize information in a picture, list, table, graph, or diagram

Materials

- *Dot Polygons* (page 109; dots.pdf)
- *Dot to Dot* (dottodot.pdf)
- *Student Response Form* (page 132; studentresponse.pdf) *(optional)*

Activate

1. Distribute *Dot to Dot* to each student. Have students find the area of each irregular figure.

2. Have students identify two different ways in which they might calculate the area of the figures.

3. Have students share one figure and its area. Ask if anyone found a different answer or did it differently. Have them share the strategy and solution. Continue until all the areas are identified.

Solve

1. Distribute copies of *Dot Polygons* to students. Have students work alone, in pairs, or in small groups.

2. As students work, you may need to help some students recognize that when breaking up the figure, they must use the lattice points as a vertex.

Debrief

1. What strategies did you find helpful when finding the areas?

2. What shapes were easier to break apart? Why?

Differentiate △

Have students investigate how they might use the number of dots in the figure to help determine the area of the irregular figures.

George likes working with small details.

- How might he box in the following figures to determine the area?
- Find the area for each figure.

 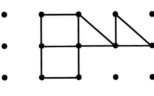 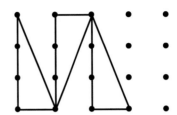

Arthur is finding the area of the two shapes below. Help him find the areas by using your knowledge of the relationship between rectangles and triangles.

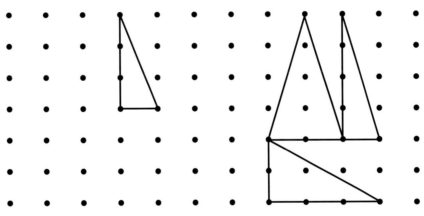

Ella is designing a geometric sculpture for her bedroom. She has 12 sq. ft. of space on the wall where she plans to hang it. Find the area of her sculpture and tell her whether she has enough space on the wall.

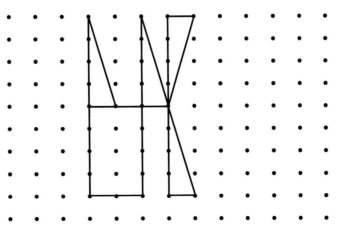

Nets

Standards

- Understands the relationship between two- and three-dimensional representations of a figure
- Understands formulas for finding measures

Overview

Students find the surface area of three-dimensional shapes.

Problem-Solving Strategies

- Act it out or use manipulatives
- Work backward

Materials

- *Nets* (page 111; nets.pdf)
- boxes of varying sizes, one per student
- rulers
- graph paper (graphpaper.pdf) *(optional)*
- scissors *(optional)*
- *Student Response Form* (page 132; studentresponse.pdf) *(optional)*

Activate

1. Distribute a box to each student. Have students examine their boxes and think about how they might determine how much cardboard was used to make the box.

2. Ask students to write a series of steps that they could use to figure out their box's surface area. Have students exchange their steps with another student and follow each of the steps as listed to find the surface area of the box.

3. Have volunteers report how successfully they were able to follow the steps and find the surface area of the box. Ask if anyone did it differently, and if so, have those students share their list of steps.

4. If students do not suggest unfolding the box into a two-dimensional net, show them how this can be done and discuss how this could be helpful in finding the surface area of a solid figure.

Solve

1. Distribute copies of *Nets* to students. Have students work alone, in pairs, or in small groups.

2. As students work, suggest they predict what the most efficient packaging dimensions would be before they calculate the surface area.

Debrief

1. What strategies did you find helpful when finding the surface area?

2. Why do you think the units for surface area are square units?

Differentiate ○

Have students sketch the two-dimensional nets on graph paper, cut them out, and fold the nets into the three-dimensional objects.

Christy is working in a gift-wrapping booth at the school fair. She discovered she could fold the following two-dimensional net into a covered box. She thinks there are other nets that can be folded into a covered box. Draw at least two other nets that can be folded into covered boxes without any overlapping faces.

Parker wants to make a rectangular prism that is 12 inches long, 4 inches wide, and 4 inches high.
- Draw a two-dimensional net for this prism.
- Find the surface area for the rectangular prism.

Alex calculated the surface area of a rectangular prism to be 190 sq. in. If the area of the base is 50 sq. in., what might the dimensions of the rectangular prism be?

Packaging Candy

Standard

Understands the relationships among linear dimensions, area, and volume, and the corresponding uses of units, square units, and cubic units of measure

Overview

Students find the volume of rectangular prisms.

Problem-Solving Strategies

- Act it out or use manipulatives
- Work backward

Materials

- *Packaging Candy* (page 113; candy.pdf)
- centimeter cubes
- scissors *(optional)*
- graph paper (graphpaper.pdf) *(optional)*
- boxes of various sizes, one per group
- *Student Response Form* (page 132; studentresponse.pdf) *(optional)*

Activate

1. Ask students to tell you what they know about volume.

2. Display a box for the class and ask students how many dimensions the box has.

3. Ask students how they might determine the number of cubes that could fit in the box. Have them think about a strategy, then share their strategy with a partner. Ask students to share the strategies they think will help them calculate the volume of the box. Record those strategies on the board.

4. Distribute boxes and cubes to small groups of students. Have students determine how many cubes will fit into their box.

5. After groups complete the task, have students share the strategies they used to calculate the volume. Ask several groups to demonstrate how they determined the volume of their boxes.

Solve

1. Distribute copies of *Packaging Candy* to students. Have students work alone, in pairs, or in small groups.

2. As students work on the problems, suggest that they predict what the most efficient packaging dimensions would be before they calculate the volume.

Debrief

1. What strategies did you find helpful when finding the volume?

2. Why do you think the units for volume are cubic units?

Differentiate ⬤

Have students draw nets on graph paper, cut them out, and fold the nets into the three-dimensional objects. Students can fill them with cubes, then complete the calculations.

The Sweet Squares Candy Company wants to package its candy squares using the least amount of cardboard as possible. If they plan on putting 36 candies in the shape of one-inch cubes in each package, what are possible box sizes? List at least three different ways they might box the candy. (Assume all measurements are whole numbers.)

Sugary Sweets is designing packages for their candy. If they put 48 candies in the shape of one-inch cubes in a box, what is the most efficient box they can design? Justify your response.

The Sweet Tooth candy store stored 496 cubic inches of candy in one box. What might the dimensions of the container be? If there are multiple possibilities, list at least four of them.

Packaging Candy

Polygons on the Plane

Standard

Uses the rectangular coordinate system to model and solve problems

Overview

Students plot polygons on the coordinate plane and find the length of a side or the distance between two vertices.

Problem-Solving Strategies

- Act it out or use manipulatives
- Organize information in a picture, list, table, graph, or diagram

Materials

- *Polygons on the Plane* (page 115; polygonsplane.pdf)
- *Plotted Polygon* (plottedpolygon.pdf)
- graph paper (graphpaper.pdf)
- *Student Response Form* (page 132; studentresponse.pdf) *(optional)*

Activate

1. Distribute *Plotted Polygon*. Have students find the length of each side on the square. *(6 units)* Then have students find the perimeter of the square. *(24 units)*

2. As students work, observe students who include absolute value notation as they set up their computation.

3. Ask several students to share their solutions for the distance between each vertex and the perimeter.

Solve

1. Distribute copies of *Polygons on the Plane* to students. Have students work alone, in pairs, or in small groups.

2. As students work on the problems, observe how they determine the side lengths of the polygons.

3. Ask clarifying questions to determine whether students recognize that they can find the side lengths by subtracting first coordinates or second coordinates.

Debrief

1. How did you decide whether to add or subtract the coordinates?

2. Did it make sense to include absolute value? Why or why not?

3. Was it possible to get a negative length?

Differentiate △

Give some of the coordinates for a polygon and have students find the missing coordinates to complete the figure.

Clint noticed that the corn maze at a local farm was laid out in the shape of a rectangle. He and his friends decided to walk the perimeter of the maze. Label each vertex. How far did the boys walk?

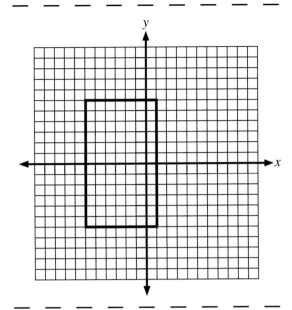

Kyle and some friends are staking out their camp area at a local campsite. They decide to put stakes at markers (9, 3), (9, –8), (–6, –8), and (–6, 3). Graph the coordinates and determine the length of each side of the camping area.

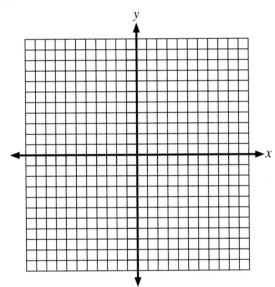

Marisol is helping her grandmother plan a rectangular play area for a day care center. She drew a scale model for her grandmother on graph paper but forgot to include the coordinate points. If the perimeter of the play area measures 48 yards, what could the coordinates for the vertices be?

Measurement Stories

Standard

Selects and uses appropriate units to find measurements for real-world problems

Overview

Students identify appropriate units of measure in real-world situations.

Problem-Solving Strategies

- Guess and check
- Use logical reasoning
- Work backward

Materials

- *Measurement Stories* (page 117; measurement.pdf)
- rulers
- meter sticks
- measuring tapes
- *Student Response Form* (page 132; studentresponse.pdf) *(optional)*

Activate

1. Ask students what they know about units of measure that are used for lengths or distances.

2. Have students measure a variety of distances in and around the classroom, such as the length and width of the room, the height of the door frame, or the length of the board.

3. Have students record the distances in an organized list or chart.

4. Discuss the various units of measure students used. Ask students to identify other units of measure they might use in various circumstances.

5. Ask students *Which is longer—one mile or one kilometer?* Have them explain their reasoning.

Solve

1. Distribute copies of *Measurement Stories* to students. Have students work alone, in pairs, or in small groups.

2. Ask clarifying questions to prompt students to explain the units of measure they are choosing.

Debrief

1. How did you decide which values went with each circumstance?

2. How did you convert between different units of measure?

Differentiate

Some students will benefit from extra practice measuring various distances. It is helpful to discuss the necessity of when accuracy is crucial versus when a close approximation is acceptable.

Complete the story using the numbers in the box below so the story makes sense.

2 ft. 11 in.	90	8.65	1979	28 ft. 4.5 in.

Lee Redmond has not cut her nails since _____. She has grown and carefully manicured them to reach a total length of _____ m, which is equivalent to _____. Her longest nail is on her right thumb. It measures _____ cm, which is _____.

Complete the story using the numbers in the box below so the story makes sence. You may use one number twice.

28.7	3.5	50	5	308	2002	13	19.7

Spike, an Asian elephant in the Calgary Zoo, was _____ pounds at birth. He recently had the largest dental caps ever made placed on his cracked tusks. The caps measure _____ cm (_____ in.) long. They are _____ cm (_____ in.) in diameter and weigh _____ kg (_____ lbs.). The operation took _____ hours and was performed on Independence Day in _____.

Complete the story using the numbers in the box below so the story makes sense.

1.5×10^8	3×10^5	9.3×10^7

A light-year is a unit of distance. It is the distance that light travels in one year. Light moves at a velocity of about _____ km per second. In the universe, the kilometer is just too small to be a useful unit of measure. Scientists use the Astronomical Unit (AU) to measure distances in the solar system. It is approximately _____ km from Earth to the sun, which is about _____ miles.

Center It

Standard

Understands basic characteristics of measures of central tendency

Overview

Students decide which measure of central tendency is most appropriate in contextual situations.

Problem-Solving Strategies

- Find information in a picture, list, table, graph, or diagram
- Organize information in a picture, list, table, graph, or diagram

Materials

- *Center It* (page 119; centerit.pdf))
- easel-sized graph paper *(optional)*
- *Student Response Form* (page 132; studentresponse.pdf) *(optional)*

Activate

1. Write the following data set on the board: 67, 89, 95, 74, 83, 89, 68, 68, 83, 83, 83.

2. Have students examine the list and ask *Which grade would you like to share with your parents—the mean (80.2), median (83), or mode (83)?* Tell the students they must justify their answers.

3. Discuss the differences among mean, median, and mode. Include the term *measures of central tendency* when discussing these terms.

Solve

1. Distribute copies of *Center It* to students. Have students work alone, in pairs, or in small groups.

2. As students are working, walk around to each group and ask questions, such as *How might you identify the median? How many data elements do you have? Can you tell me what the mode is? What does the mean actually represent?*

3. Have students share their solutions. Ask if anyone got a different answer or did it differently. Record answers on the board to encourage discussion.

Debrief

1. What information did the mean provide? The mode? The median?

2. Give an example of the use of median in the real world.

Differentiate ☆

Provide English language learners with the vocabulary before teaching the lesson. Make a word wall with the terms *mean*, *median*, and *mode*, and clearly define and provide examples of each term.

Your basketball coach was bragging to an opposing team's coach about the average number of free-throw shots your team has made this season. If the number of free-throw shots made per player included 17, 13, 7, 9, 10, 15, 14, 11, 16, 18, and 20, should your coach use the mean, median, or mode to make your team look better? Justify your response.

Your class is competing in the school fitness challenge. You have been recording the number of push-ups each person in the class can do in a one-minute session each day. Today, your class's data included 13, 15, 16, 17, 17, 8, 19, 9, 17, 15, 11, 13, 12, 16, 11, and 13. Should you report the mean, median, or mode of your data to make your class look better? Justify your response.

Your school's baseball team is traveling to the rival school to compete. In preparation, your coach told the team that the rival school's team batting average is over .300. Did your coach report the team's batting average using the mean, median, or mode, if the individual batting averages are .344, .251, .309, .087, .314, .246, .292, .248, .179, .280, .272, and .314? Justify your response.

Change It

Standard

Understands basic characteristics of measures of central tendency

Overview

Students examine the effect of adding or subtracting data from a data list in which the mean, median, and mode have already been calculated.

Problem-Solving Strategies

- Find information in a picture, list, table, graph, or diagram
- Use logical reasoning

Materials

- *Change It* (page 121; changeit.pdf)
- calculators (*optional*)
- linking cubes (*optional*)
- *Student Response Form* (page 132; studentresponse.pdf) (*optional*)

Activate

1. Have students calculate the mean (*26*), median (*24.5*), and mode (*21*) for the following data set: 25, 29, 18, 34, 22, 21, 31, 32, 35, 20, 21, 24.

2. Add new data to the list. Ask students to interpret the impact on the measures of central tendency. Have students share their thinking.

3. Ask students to discuss how they might add data without changing the median.

Solve

1. Distribute copies of *Change It* to students. Have students work alone, in pairs, or in small groups.

2. As students are working, walk around to each group and ask questions, such as *What measure of central tendency balances the data among the number of data included? What measure of central tendency can be used to divide the data into two equivalent groups? How many modes might a data set have?*

3. Ask students to share their solutions. Ask if anyone got a different answer or did it differently. Be sure to record all answers on the board to encourage discussion.

Debrief

1. Why do you think some costs are represented by the median rather than the mean?

2. What information does the mode provide?

Differentiate ⬤ ☆

Use linking cubes to model situations that may have 6 pieces of data which differ, such as 4, 6, 2, 3, 5, 6. To find the mean, students then must evenly distribute the cubes so all stacks of cubes have the same amount.

Carrie averages about 8 hours of babysitting each week. She babysits 5 days a week for a family whose children need care after school. If she works the same number of hours each day, how many hours does she babysit each day?

Jerome recorded the number of hours he spends playing or practicing hockey each week. What would happen to the mean if next week he played hockey for 6 hours on Monday, but played or practiced his usual number of hours on the other days?

Day	Hours
Monday	5
Tuesday	4
Wednesday	7
Thursday	2
Friday	3
Saturday	6
Sunday	4

Ashley made a list of the amount of time she spent reading one week. How many hours might she spend reading on one other day so that the median does not change?

Day	Hours
Monday	3
Tuesday	1
Wednesday	2
Thursday	4
Friday	3
Saturday	2
Sunday	3

Mean It

Standard

Understands basic characteristics of measures of central tendency

Overview

Students find the mean of a data set or, given the mean, calculate missing values from a list of data elements.

Problem-Solving Strategies

- Count, compute, or write an equation
- Work backward

Materials

- *Mean It* (page 123; meanit.pdf)
- snap cubes
- *Student Response Form* (page 132; studentresponse.pdf) *(optional)*

Activate

1. Distribute a large number of snap cubes to pairs of students. Have them stack the cubes so there are stacks with 4, 7, 8, 2, and 3 cubes in the stacks.

2. Ask them to find out how many cubes would be in a sixth stack if the mean for the data set is 6. (*12*) Ask students to think about possible strategies they can use to find the missing value.

3. Ask students to share their strategies for finding the value.

4. Ask if anyone got a different answer or used a different strategy. If so, have them share their work.

Solve

1. Distribute copies of *Mean It* to students. Have students work alone, in pairs, or in small groups.

2. As students are working, walk around to each group and ask questions, such as *How did you find the mean? What did the total represent before you divided by the number of data elements? How does that sum relate to finding the missing value when you know the mean?*

3. Ask students to share their solutions. Ask if anyone got a different answer or did it differently. Be sure to record all answers on the board to encourage discussion.

Debrief

1. What information did the mean tell you?

2. Why do you think that the mean was most commonly reported as the average?

Differentiate ◯ ▢ △ ☆

Have students complete an exit card task, such as *How does the mean differ from the median?*

Joanna received her lap times for the past week in swimming. What is the average amount of time it took her to swim a lap?

Day	Time
Monday	4 min.
Tuesday	5 min.
Wednesday	5 min.
Thursday	4 min.
Friday	4 min.
Saturday	5 min.
Sunday	3 min.

Shannon exercised for different amounts of time every day. What is her average exercise time this week? How much time should she exercise the next day to achieve an average of 45 minutes per day?

Day	Time
Monday	45 min.
Tuesday	30 min.
Wednesday	55 min.
Thursday	25 min.
Friday	60 min.
Saturday	35 min.
Sunday	25 min.

Yolanda wanted to run an average of 60 minutes each day. Find the mean for the six days. How many minutes does she need to run on the seventh day to average 60 minutes of running each day?

Day	Time
Monday	38 min.
Tuesday	55 min.
Wednesday	43 min.
Thursday	45 min.
Friday	60 min.
Saturday	55 min.
Sunday	?

Mean It

Statistical Questions

Standard

Uses data and statistical measures for a variety of purposes

Overview

Students design questions for statistical analysis.

Problem-Solving Strategy

Use logical reasoning

Materials

- *Statistical Questions* (page 125; statquestions.pdf)
- *Student Response Form* (page 132; studentresponse.pdf) *(optional)*

Activate

1. Ask students what they think it means to ask a statistical question.

2. Ask students to brainstorm a list of statistical questions. Write the list on the board.

Solve

1. Distribute copies of *Statistical Questions* to students. Have students work alone, in pairs, or in small groups.

2. As students are working, walk around to each group and ask questions, such as *Does your question look for a general response or are you just looking to understand what one person prefers? Does the answer to the question allow you to make a general statement about the group that is responding?*

3. Ask students to share a solution. Ask if anyone got a different answer or did it differently. Be sure to record all answers on the board to encourage discussion.

Debrief

1. What was a statistical question?

2. How did you decide on the statistical question?

Differentiate ⬤ ▢ △ ☆

Have students complete an exit card, such as *Write a statistical question you might pose to your classmates if you were doing a survey.*

Statistical Questions

Jenna is interested in finding out something all her classmates like to do. She has decided to make up a survey about her friends' favorite books. What question might Jenna pose?

Statistical Questions

Jack took a survey of his friends. He made a circle graph to show the results. What do you think his question might have been?

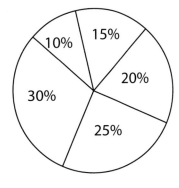

You are curious about how many of your friends like the school lunches. How might you conduct a survey? What question will you ask? How will you gather your data? How can you be sure your data is not biased? How might you represent your data?

Stem-and-Leaf

Standards

- Reads and interprets data in plots
- Organizes and displays data using plots

Overview

Students interpret stem-and-leaf plots.

Problem-Solving Strategies

- Find information in a picture, list, table, graph, or diagram
- Use logical reasoning

Materials

- *Stem-and-Leaf* (page 127; stemleaf.pdf)
- graphing calculators (*optional*)
- *Student Response Form* (page 132; studentresponse.pdf) (*optional*)

Activate

1. Ask students what they know about stem-and-leaf plots.

2. Write the following numbers on the board: 8, 13, 18, 21, 25, 32. Show students how to make a stem-and-leaf plot with the data set.

0	8	
1	3	8
2	1	5
3	2	

3. Ask students to determine the median and mode of the data set.

4. Have students share how the median and mode were found.

5. Ask if anyone did it differently or got a different answer. Discuss any misconceptions.

Solve

1. Distribute copies of *Stem-and-Leaf* to students. Have students work alone, in pairs, or in small groups.

2. As students are working, walk around to each group and ask questions such as *What does the data tell you? What does the median tell you? How might you determine the median? Mode? Mean?*

3. Ask students to share their solutions. Ask if anyone got a different answer or did it differently. Record answers on the board.

Debrief

1. What information did the stem-and-leaf plots tell you?

2. How was the data ordered on the stem-and-leaf plots? Did the order matter?

3. What are some other statistical measures that describe a data set?

Differentiate △

Have students make a histogram of the data. Encourage them to explore the statistical functions on a graphing calculator.

Mohammed analyzed the data from a survey he took. Draw a stem-and-leaf plot that represents the data.

63   35   46   78   13   16   17   19   23   19   31   19

Carly examined the stem-and-leaf plot below. She wanted to know what the median score was. Find the median.

1	0 1 3 5 6 6 7 9 9 9
2	0 3 3 4 4 4 5 7
3	0 1 1 3 4 5
4	3 6 6 8 8 9
5	1 1 1
6	1 3 4
7	8
8	
9	5 5 6 7 8 8 9 9

Alejandro recorded the number of push-ups students in his class did and compared them with the number done by students in his brother's class. Which class has a better push-up average? Explain how you know.

Joaquin's Class									Alejandro's Class									
8 8 7 6 6 5 5 3								0	0 1 3 5 6 6 7 9 9 9									
7 7 7 6 3 3 3 2								1	3 6 6 8 8 9									
4 4 1								2	0 3 3 4 4 4									
4 3 2 1 0								3	0 1 1 2									
1 1								4										
1								5										
								6	2									

Line Plots

Standard

Organizes and displays data using plots

Overview

Students make a line plot and interpret data from it.

Problem-Solving Strategies

- Act it out or use manipulatives
- Find information in a picture, list, table, graph, or diagram
- Organize information in a picture, list, table, graph, or diagram

Materials

- *Line Plots* (page 129; lineplots.pdf)
- blocks or colored discs *(optional)*
- *Student Response Form* (page 132; studentresponse.pdf) *(optional)*

Activate

1. Ask students what they know about line plots.
2. Ask students how many pets they own. Record their answers on the board.
3. Show students how to create a line plot with the pet data.
4. Discuss the pros and cons of using line plots.

Solve

1. Distribute copies of *Line Plots* to students. Have students work alone, in pairs, or in small groups.
2. As students are working, walk around and ask questions, such as *What can you say about the shape? Can you tell where the median is by inspection? Tell me what you think the mode is. How might you use this data to find the mean? Do you think the mean is greater than, less than, or equal to the median?*
3. Have students share their strategies and solutions. Record answers on the board and encourage discussion.

Debrief

1. What information did you see in a line plot?
2. What are some reasons you might use a line plot?

Differentiate ◯☆

Many students benefit from building a line plot using blocks or colored discs, which can be manipulated easily. Students can move the blocks or discs around to help them find the mean.

Gary asked his classmates how many books they read last year. He made the following line plot of his data.

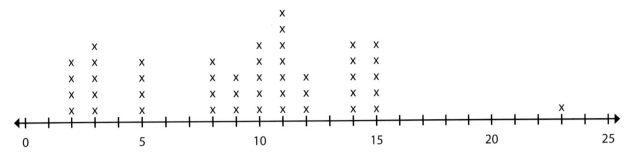

- How many people did Gary ask?
- What is the mode of the data set?
- What is the median of the data set?

Lily asked the spectators at a school basketball game how old they were. She wrote her data in the chart below. Make a line plot to display her data.

7	10	36	29	7	35	8	12	18	37	7	10
36	29	7	35	8	12	18	37	36	10	36	29
7	35	8	12	18	37	7	36	36	9	9	9

The following data show how many sit-ups the 6th grade students completed in the school olympics. Make a line plot for the data. What is the mean, median, and mode for the data set?

30, 37, 36, 34, 49, 35, 40, 47, 47, 39, 54, 47, 48, 54, 50, 35, 40, 38, 47, 48, 34, 40, 46, 49, 47, 35

Box and Whiskers

Standard

Organizes and displays data using plots

Overview

Students make box-and-whisker plots and interpret data from them.

Problem-Solving Strategies

- Find information in a picture, list, table, graph, or diagram
- Organize information in a picture, list, table, graph, or diagram

Materials

- *Box and Whiskers* (page 131; boxwhiskers.pdf)
- *Student Response Form* (page 132; studentresponse.pdf) *(optional)*

Activate

1. Ask students what they can say about a data set if they only know the minimum and the maximum values.

2. Draw a box-and-whisker plot on the board. Ask students to make a list of observations. Have students discuss their lists with a partner.

3. After each pair of students has shared their observations, discuss the pros and cons of using box-and-whisker plots.

Solve

1. Distribute copies of *Box and Whiskers* to students. Have students work alone, in pairs, or in small groups.

2. As students are working, walk around and ask questions, such as *How might you identify the median? How many data elements do you have? What is the mode? What does the box represent?*

3. Have students share their solutions. Ask if anyone got a different answer or did it differently. Be sure to record all answers on the board to encourage discussion.

Debrief

1. What information do you not get from a box-and-whisker plot?

2. What are some of the reasons you might use a box-and-whisker plot?

Differentiate

Many students benefit from building a box-and-whisker plot using different lengths of licorice, coffee stirrers, or straws. Have students organize their information in a table or chart before creating a box-and-whisker plot.

Juan was asked to examine the box-and-whisker plot below.

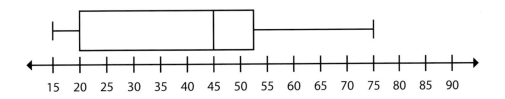

- What is the median?
- What is the range?

Luke was asked to draw a box-and-whisker plot to represent the following data. Graph the data. Identify the median and range.

36, 12, 37, 9, 7, 12, 35, 8, 35, 29, 7, 9, 18, 36, 10, 29, 12, 36, 7, 8, 35, 7, 8, 36, 18, 7, 8, 37, 7, 37, 9, 10, 36

Lindsay examined the following box-and-whisker plot, which had no explanation or context with it. How many more data are there between 38 to 82 then there are between 82 to 92? How do you know? Justify your response.

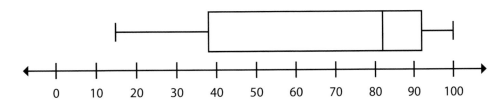

Name: _____ Date: _____

Student Response Form

Problem:

(glue your problem here)

My Work and Illustrations:
(picture, table, list, graph)

My Solution:

My Explanation:

_____

_____

_____

Name: _____ Date: _____

Observation Form

	Criteria	Notes
Communication	Variety of Methods (written, oral, etc.)	
	Interaction with Peers • Are they on topic? • Are they respectful?	
	Teacher/Student How does student respond to teacher inquiry?	
	Clarifying Questions What types of questions are being asked within the group?	
	Makes Connections • Problem to other math concepts • Problem to real world • Problem to other content	
Problem Solving	Chooses Appropriate Strategies • What strategy did they choose? • Is the strategy efficient?	
	Reasonableness Does the answer make sense numerically and contextually?	
	Confidence in approaching problem	
Reasoning and Proof	Defend and Justify Can they mathematically defend and justify their thinking?	
	Inductive vs. Deductive • Inductive—looks for patterns and makes generalizations • Deductive—makes logical arguments, draws conclusions, applies generalizations to specific situations	
Content	Accuracy • How accurate is the work being done? • Are there any particular mistakes being made?	

Record-Keeping Chart

Use this chart to record the problems that were completed. Record the name of the lesson and the date when the appropriate level was completed.

Name: _____

Lesson	⬤ Date Completed	⬛ Date Completed	▲ Date Completed

Answer Key

How Many Prefer? (page 33)

●
Bubblegum	6	12	18	24	54
Chocolate Chip	4	8	12	16	36

54 sixth graders

■ 84 players; 30%

▲ 90 students; 60 students

Survey Results (page 35)

●
580	580	580

1,160 students

■ 4,980 dogs; 7,470 dogs

▲ 3:2

Best Buys (page 37)

● Jack should buy Jumbo Chips.

■ Bike A

▲ Buying 11 DVDs for $49.99 and the remaining two at $4.99 each is the best buy.

Paint Colors (page 39)

●

Tommy	
Green Paint	White Paint
5	7
10	14
15	21
20	28

Renee	
Green Paint	White Paint
7	12
14	24
21	36
28	48

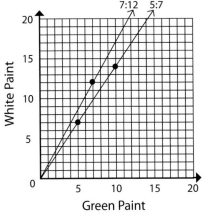

■ Answers will vary.

▲ darker red

Mixing It Up (page 41)

● 144 beads

■ 50 cards

▲ 105 marbles

Percent Tables (page 43)

●
Discount	Discount Percent
$120	100%
$60	50%
$30	25%

$30

■
Discount	Discount Percent
$499	100%
$249.50	50%
$124.75	25%
$49.90	10%

$199.60; $299.40

▲ $225.25

How Many Groups? (page 45)

● 2 friends

■ 7 people

▲ No; She needs to buy 2.5 more pints.

Identical Groups (page 47)

● 60 bracelets; 9 black beads; 10 silver beads

■ 8 piles; 2 quarters; 4 dimes; 5 nickels

▲ 18 baskets; 4 roses; 3 tulips; 2 carnations

Are We in Sync? (page 49)

● Friday, January 5

■ 5 times; 25 times

▲ 9:42 AM; 8 times

Factors or Multiples? (page 51)

● 8 packages of stamps and 5 packages of cards

■ 9:35 AM; 5 times

▲ 8 backpacks

Answer Key (cont.)

What's My Value? (page 53)

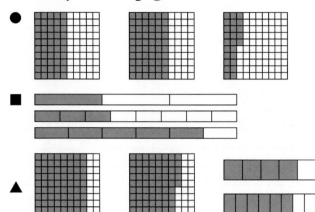

Greater or Less Than Zero? (page 55)

● –30 ft.

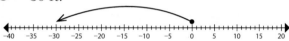

■ Dropped 15°F

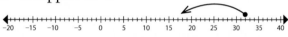

▲ 7 blocks

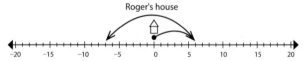

Integer Values (page 57)

● H, E, F, I, A, C, D, G, B; 7; –15

■ –35, –21, –18, –13, –11, –9, –4, –3, –1

▲ No

Opposites Attract (page 59)

● 0

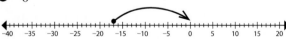

■ 8 + 18 = 26 or 8 + –18 = –10

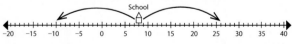

▲

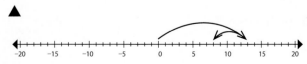

Coordinate Graphing (page 61)

● 16 blocks

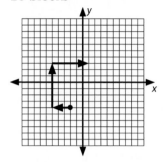

■ 21 miles; Answers will vary.

▲ Answers will vary.

Computing with Integers (page 63)

● –3

■ –6 + 0

▲ 17 + –5 + –5 + –5 = 2

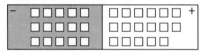

What's My Number? (page 65)

● 600.024; 4.096; 9.014

■ 6,000.027; 80,000.12; 70.0305

▲ 50.9206; 429.86454; Answers will vary.

Garden Areas (page 67)

● A = 36 sq. meters

■ 3.45 m

▲ Explanations will vary.

Methods of Multiplying (page 69)

● 2,254 stickers

■ 7,236 cards

▲ Answers will vary.; 2,496 marbles

Bank On It (page 71)

● $0.50

■ Yes; $5 left over

▲ The bank is in the red; –$62.45 + $34.50 = –$27.95

Answer Key *(cont.)*

Exponentials (page 73)

- ● 8^5; 7^{12}; 4^3
- ■ 3^4; 12^{11}; b^3; Explanations will vary.
- ▲ $d \times d \times d \times d \times d \times d$; b^4; $(ab)^5$; $(cd)^7$

Balancing Act (page 75)

- ● 1 sphere
- ■ 2 trapezoids
- ▲ 4

Evaluate Me (page 77)

- ● 22
- ■ −33
- ▲ He is incorrect. $(-3)(5) + 5(3 \times 5 + 2) =$ $-15 + 5(17) = 70$

Variable Value (page 79)

- ● a. $b = 3$
  b. $c = 9$
  c. $d = 13$
- ■ a. $d = 6$
  b. $b = 0$
  c. $e = -14$
- ▲ a. $m = 5\frac{1}{4}$
  b. $y = 6.25$
  c. $b = 33\frac{1}{2}$

Systems of Equations (page 81)

- ● $\triangle = 13, \bigcirc = 22, \square = 19$
- ■ $\triangle = 11, \bigcirc = 15, \square = 7$
- ▲ $\triangle = 15, \bigcirc = 12, \square = 6$

Heads and Feet (page 83)

- ● 23 hens and 17 goats
- ■ 31 zebras and 15 ostriches
- ▲ 48 cows and 17 roosters

How Much Money? (page 85)

- ● 2 one-dollar bills, 2 five-dollar bills, and 11 ten-dollar bills
- ■ Answers will vary.
- ▲ 23 adults and 73 students

My Equation Is (page 87)

- ● Tyler: 62 minutes, Randi: 90 minutes; $t + t + 28 = 152$
- ■ Penny: 231 minutes, Desmond: 353 minutes; $p + p + 122 = 584$
- ▲ Scott: 43 minutes, Damien: 167 minutes; $s + s + 124 = 210$

Arithmetic Sequences (page 89)

- ● 35, 42, 49; $7n$
  17, 19, 21; $2n + 1$
- ■ 22, 27, 32; $5n - 3$
  27, 31, 35; $4n + 7$
- ▲ 28, 34, 40; $6n - 2$
  −13, −16, −19; $-3n + 2$

Going the Distance (page 91)

- ● 150 miles
- ■

Speed	Time
40 mph	3.4 hrs.
50 mph	2.7 hrs.
60 mph	2.3 hrs.
65 mph	2.1 hrs.

Explanations will vary

▲

Time	Speed
$3\frac{1}{2}$ hrs.	47 mph
4 hrs. 20 mins.	38 mph
$2\frac{1}{2}$ hrs.	66 mph

$165 \div h = r$

Answer Key *(cont.)*

Arithmetic and Geometric Sequences (page 93)

- ● arithmetic; 31; 59

  geometric; 64; 8,192

- ■ geometric; 1; 8

  arithmetic; 18; 36

- ▲ geometric; 18; 54; 162

  arithmetic; 8; 13; 21; sum of the previous two numbers

Various and Sundry Patterns (page 95)

- ● 6; 11; $n + 1$
- ■ 28; 55; add the figure number and the number of hexagons in the previous figure
- ▲ 49; 225; n^2

Expressly What? (page 97)

- ● $(x + 8) \times 4 = 56$
- ■ $[(x \times 2) - 8] \times 6 = 36$
- ▲ $(x - 10) \times 5 + 4 = 19$

Simplify Me (page 99)

- ● a. 35   b. −9
- ■ a. −15   b. −20
- ▲ a. $28\frac{3}{4}$ b. 8

How Far Did I Go? (page 101)

- ● 130 miles; $227\frac{1}{2}$ miles; 325 miles; $422\frac{1}{2}$ miles; $d = 65t$
- ■ 75 mph; $d = 75t$
- ▲ Problems will vary.

Equivalences (page 103)

- ● No; Kelly didn't multiply the 8 and the 4 before dividing by 8.; 32
- ■ Raoul didn't divide 1 by -4 before adding.; $-23\frac{1}{4}$
- ▲ Expressions will vary.

Quadrilaterals and Triangles (page 105)

- ● Strategies will vary; triangle: 15 sq. units; trapezoid: 21 sq. units; parallelogram: 35 sq. units
- ■ Predictions will vary; $A = 30$ sq. cm; $A = 10$ sq. cm; $A = 42$ sq. cm
- ▲ Predictions will vary; $A = 6$ sq. in.; $A = 12$ sq. in.; $A = 9$ sq. in.

Boxy Areas (page 107)

- ● 55 sq. ft.
- ■ 480 plants
- ▲ 288 sq. cm

Dot Polygons (page 109)

- ● Drawings will vary; $A = 3\frac{1}{2}$ sq. units; $A = 3$ sq. units; $A = 4\frac{1}{2}$ sq. units
- ■ $A = 1\frac{1}{2}$ sq. units; $A = 9$ sq. units
- ▲ $A = 16$ sq. ft.; She does not have enough space.

Nets (page 111)

- ● Nets will vary, but can include:

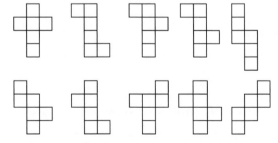

- ■

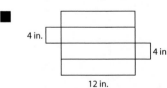

  $SA = 224$ sq. in.
- ▲ Answers will vary.

Answer Key *(cont.)*

Packaging Candy (page 113)

- Answers will vary but may include:

  2 in. × 2 in. × 9 in.; 2 in. × 3 in. × 6 in.;
  4 in. × 3 in. × 3 in.; 18 in. × 2 in. × 1 in.;
  1 in. × 1 in. × 36 in.; 6 in. × 6 in. × 1 in.;
  12 in. × 3 in. × 1 in.

- ■ 4 in. × 4 in. × 3 in.

- ▲ Answers will vary but may include:

  1 in. × 16 in. × 31 in.;
  2 in. × 8 in. × 31 in.;
  4 in. × 4 in. × 31 in.; 1 in. × 2 in. × 248 in.

Polygons on the Plane (page 115)

- (1, 6), (1, –6), (–6, 6), (–6, –6); 38 units

- ■

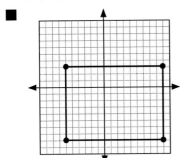

  length: 15 units, width: 11 units

- ▲ Answers will vary.

Measurement Stories (page 117)

- 1979; 8.65; 28 ft. 4.5 in.; 90; 2 ft. 11 in.
- ■ 308; 50; 19.7; 13; 5; 13; 28.7; 3.5; 2002
- ▲ 3×10^5; 1.5×10^8; 9.3×10^7

Center It (page 119)

- median
- ■ mode of 17
- ▲ mode

Change It (page 121)

- 1 hour 36 minutes per day
- ■ The mean would increase by 0.1.
- ▲ Answers will vary.

Mean It (page 123)

- $4\frac{2}{7}$ minutes
- ■ $39\frac{2}{7}$ minutes; 85 minutes
- ▲ $49\frac{1}{3}$ minutes; 124 minutes

Statistical Questions (page 125)

- Questions will vary.
- ■ Questions will vary.
- ▲ Answers will vary.

Stem-and-Leaf (page 127)

-
1	3 6 7 9 9 9
2	3
3	1 5
4	6
5	
6	3
7	8

- ■ 34

- ▲ Joaquin's class; Explanations will vary.

Line Plots (page 129)

- 46 people; 11; 10

- ■

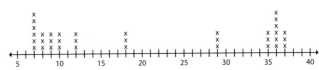

- ▲ mean = 42.5; median = 43; mode = 47

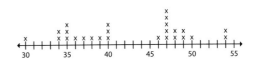

Box and Whiskers (page 131)

- 45; 60

- ■ median = 12; range = 30

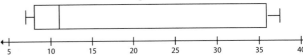

- ▲ There are the same amount of data in each quartile.

References Cited

Bright, G. W., and J. M. Joyner. 2005. *Dynamic classroom assessment: Linking mathematical understanding to instruction.* Chicago, IL: ETA Cuisenaire.

Brown, S. I., and M. I. Walter. 2005. *The art of problem posing.* Mahwah, NJ: Lawrence Earlbaum.

Cai, J. 2010. Helping elementary students become successful mathematical problem solvers. In *Teaching and learning mathematics: Translating research for elementary school teachers,* ed. D. V. Lambdin and F. K. Lester, Jr., 9–13. Reston, VA: NCTM.

D'Ambrosio, B. 2003. Teaching mathematics through problem solving: A historical perspective. In *Teaching mathematics through problem solving: Prekindergarten–Grade 6,* ed. F. K. Lester, Jr. and R. I. Charles, 37–50. Reston, VA: NCTM.

Goldenberg, E. P., N. Shteingold, and N. Feurzeig. 2003. Mathematical habits of mind for young children. In *Teaching mathematics through problem solving: Prekindergarten–Grade 6,* ed. F. K. Lester, Jr. and R. I. Charles, 51–61. Reston, VA: NCTM.

Michaels, S., C. O'Connor, and L. B. Resnick. 2008. Deliberative discourse idealized and realized: Accountable talk in the classroom and in civil life. *Studies in philosophy and education* 27 (4): 283–297.

National Center for Educational Statistics. 2010. Highlights from PISA 2009: Performance of U.S. 15-year-old students in reading, mathematics, and science literacy in an international context. http://nces.ed.gov/pubsearch/pubsinfo.asp?pubid=2011004

National Governors Association Center for Best Practices and Council of Chief State School Officers. 2010. Common core state standards. http://www.corestandards.org/the-standards.

National Mathematics Advisory Panel. 2008. *Foundations for success: The final report of the National Mathematics Advisory Panel.* Washington, DC: U.S. Department of Education.

Polya, G. 1945. *How to solve it: A new aspect of mathematical method.* Princeton, NJ: Princeton University Press.

Sylwester, R. 2003. *A biological brain in a cultural classroom.* Thousand Oaks, CA: Corwin Press.

Tomlinson, C. A. 2003. *Fulfilling the promise of the differentiated classroom: Strategies and tools for responsive teaching.* Alexandria, VA: ASCD.

Vygotsky, L. 1986. *Thought and language.* Cambridge, MA: MIT Press.

Contents of the Teacher Resource CD

Teacher Resources

Page	Resource	Filename
27–31	Common Core State Standards Correlation	ccss.pdf
N/A	NCTM Standards Correlation	nctm.pdf
N/A	TESOL Standards Correlation	tesol.pdf
N/A	McREL Standards Correlation	mcrel.pdf
132	Student Response Form	studentresponse.pdf
133	Observation Form	obs.pdf
134	Record-Keeping Chart	recordkeeping.pdf
N/A	Exit Card Template	exitcard.pdf

Lesson Resource Pages

Page	Lesson	Filename
33	How Many Prefer?	prefer.pdf
35	Survey Results	survey.pdf
37	Best Buys	bestbuys.pdf
39	Paint Colors	paintcolors.pdf
41	Mixing It Up	mixing.pdf
43	Percent Tables	percenttables.pdf
45	How Many Groups?	howmanygroups.pdf
47	Identical Groups	groups.pdf
49	Are We in Sync?	sync.pdf
51	Factors or Multiples?	factorsmultiples.pdf
53	What's My Value?	value.pdf
55	Greater or Less Than Zero?	greaterless.pdf
57	Integer Values	integervalues.pdf
59	Opposites Attract	opposites.pdf
61	Coordinate Graphing	coordinategraphing.pdf
63	Computing with Integers	computingintegers.pdf
65	What's My Number?	whatnumber.pdf
67	Garden Areas	garden.pdf
69	Methods of Multiplying	multiplying.pdf
71	Bank On It	bankonit.pdf
73	Exponentials	exponentials.pdf
75	Balancing Act	balancing.pdf

Contents of the Teacher Resource CD *(cont.)*

Lesson Resource Pages *(cont.)*

Page	Lesson	Filename
77	Evaluate Me	evaluateme.pdf
79	Variable Value	variable.pdf
81	Systems of Equations	systemsequations.pdf
83	Heads and Feet	headsfeet.pdf
85	How Much Money?	howmuchmoney.pdf
87	My Equation Is	myequation.pdf
89	Arithmetic Sequences	arithmetic.pdf
91	Going the Distance	distance.pdf
93	Arithmetic and Geometric Sequences	arithgeoseq.pdf
95	Various and Sundry Patterns	varioussundry.pdf
97	Expressly What?	expresslywhat.pdf
99	Simplify Me	simplifyme.pdf
101	How Far Did I Go?	howfar.pdf
103	Equivalences	equivalences.pdf
105	Quadrilaterals and Triangles	quadtri.pdf
107	Boxy Areas	boxyareas.pdf
109	Dot Polygons	dots.pdf
111	Nets	nets.pdf
113	Packaging Candy	candy.pdf
115	Polygons on the Plane	polygonsplane.pdf
117	Measurement Stories	measurement.pdf
119	Center It	centerit.pdf
121	Change It	changeit.pdf
123	Mean It	meanit.pdf
125	Statistical Questions	statquestions.pdf
127	Stem-and-Leaf	stemleaf.pdf
129	Line Plots	lineplots.pdf
131	Box and Whiskers	boxwhiskers.pdf

Contents of the Teacher Resource CD *(cont.)*

Additional Lesson Resources

Page	Resource	Filename
38, 46, 66, 68, 86, 88, 90, 92, 94, 100, 104, 106, 110, 112, 114	graph paper	graphpaper.pdf
50	Which One Am I?	whichone.pdf
60	Coordinate Plane	coordinateplane.pdf
62	Algebra Mat	algebramat.pdf
72	Exponents	exponents.pdf
108	Dot to Dot	dottodot.pdf
114	Plotted Polygon	plottedpolygon.pdf

Notes